The IMA Volumes
in Mathematics
and its Applications

Volume 58

Series Editors
Avner Friedman   Willard Miller, Jr.

Institute for Mathematics and its Applications
IMA

The **Institute for Mathematics and its Applications** was established by a grant from the National Science Foundation to the University of Minnesota in 1982. The IMA seeks to encourage the development and study of fresh mathematical concepts and questions of concern to the other sciences by bringing together mathematicians and scientists from diverse fields in an atmosphere that will stimulate discussion and collaboration.

The IMA Volumes are intended to involve the broader scientific community in this process.

Avner Friedman, Director
Willard Miller, Jr., Associate Director

* * * * * * * * * *

IMA ANNUAL PROGRAMS

1982–1983	Statistical and Continuum Approaches to Phase Transition
1983–1984	Mathematical Models for the Economics of Decentralized Resource Allocation
1984–1985	Continuum Physics and Partial Differential Equations
1985–1986	Stochastic Differential Equations and Their Applications
1986–1987	Scientific Computation
1987–1988	Applied Combinatorics
1988–1989	Nonlinear Waves
1989–1990	Dynamical Systems and Their Applications
1990–1991	Phase Transitions and Free Boundaries
1991–1992	Applied Linear Algebra
1992–1993	Control Theory and its Applications
1993–1994	Emerging Applications of Probability
1994–1995	Waves and Scattering
1995–1996	Mathematical Methods in Material Science

IMA SUMMER PROGRAMS

1987	Robotics
1988	Signal Processing
1989	Robustness, Diagnostics, Computing and Graphics in Statistics
1990	Radar and Sonar (June 18 - June 29)
	New Directions in Time Series Analysis (July 2 - July 27)
1991	Semiconductors
1992	Environmental Studies: Mathematical, Computational, and Statistical Analysis
1993	Modeling, Mesh Generation, and Adaptive Numerical Methods for Partial Differential Equations
1994	Molecular Biology

* * * * * * * * * *

SPRINGER LECTURE NOTES FROM THE IMA:

The Mathematics and Physics of Disordered Media
> Editors: Barry Hughes and Barry Ninham
> (Lecture Notes in Math., Volume 1035, 1983)

Orienting Polymers
> Editor: J.L. Ericksen
> (Lecture Notes in Math., Volume 1063, 1984)

New Perspectives in Thermodynamics
> Editor: James Serrin
> (Springer-Verlag, 1986)

Models of Economic Dynamics
> Editor: Hugo Sonnenschein
> (Lecture Notes in Econ., Volume 264, 1986)

W.M. Coughran, Jr.   Julian Cole
Peter Lloyd   Jacob K. White
Editors

Semiconductors
Part I

With 55 Illustrations

Springer-Verlag
New York Berlin Heidelberg London Paris
Tokyo Hong Kong Barcelona Budapest

W.M. Coughran, Jr.
AT&T Bell Laboratories
600 Mountain Ave., Rm. 2T-502
Murray Hill, NJ 07974-0636 USA

Peter Lloyd
AT&T Bell Laboratories
Technology CAD
1247 S. Cedar Crest Blvd.
Allentown, PA 18103-6265 USA

Julian Cole
Department of Mathematical Sciences
Rensselaer Polytechnic Institute
Troy, NY 12180 USA

Jacob K. White
Massachusetts Institute of Technology
Department of Electrical Engineering and
  Computer Science
50 Vassar St., Rm. 36-880
Cambridge, MA 02139 USA

Series Editors:
Avner Friedman
Willard Miller, Jr.
Institute for Mathematics and its
  Applications
University of Minnesota
Minneapolis, MN 55455 USA

Mathematics Subject Classifications (1991): 35-XX, 60-XX, 76-XX, 76P05, 81UXX, 82DXX, 35K57, 47N70, 00A71, 00A72, 81T80, 93A30, 82B40, 82C40, 82C70, 65L60, 65M60, 94CXX, 34D15, 35B25

Library of Congress Cataloging-in-Publication Data
Semiconductors / W.M. Coughran, Jr. ... [et al.].
    p.   cm. — (The IMA volumes in mathematics and its
applications ; v. 58–59)
    Includes bibliographical references and index.
    ISBN 0-387-94250-5 (v. 1 : alk. paper). — ISBN 0-387-94251-3 (v.
2 : alk. paper)
    1. Semiconductors—Mathematical models.   2. Semiconductors-
-Computer simulation.   3. Computer-aided design.   I. Coughran,
William Marvin.   II. Series.
TK7871.85.S4693   1994
621.3815'2—dc20   93-50622

Printed on acid-free paper.

© 1994 Springer-Verlag New York, Inc.
All rights reserved. This work may not be translated or copied in whole or in part without the written permission of the publisher (Springer-Verlag New York, Inc., 175 Fifth Avenue, New York, NY 10010, USA), except for brief excerpts in connection with reviews or scholarly analysis. Use in connection with any form of information storage and retrieval, electronic adaptation, computer software, or by similar or dissimilar methodology now known or hereafter developed is forbidden.
The use of general descriptive names, trade names, trademarks, etc., in this publication, even if the former are not especially identified, is not to be taken as a sign that such names, as understood by the Trade Marks and Merchandise Marks Act, may accordingly be used freely by anyone.
Permission to photocopy for internal or personal use, or the internal or personal use of specific clients, is granted by Springer-Verlag, Inc., for libraries registered with the Copyright Clearance Center (CCC), provided that the base fee of $5.00 per copy, plus $0.20 per page, is paid directly to CCC, 222 Rosewood Drive, Danvers, MA 01923, USA. Special requests should be addressed directly to Springer-Verlag New York, 175 Fifth Avenue, New York, NY 10010, USA.
ISBN 0-387-94250-5/1994 $5.00 + 0.20

Production managed by Laura Carlson; manufacturing supervised by Jacqui Ashri.
Camera-ready copy prepared by the IMA.
Printed and bound by Edwards Brothers, Inc., Ann Arbor, MI.
Printed in the United States of America.

9 8 7 6 5 4 3 2 1

ISBN 0-387-94250-5 Springer-Verlag New York Berlin Heidelberg
ISBN 3-540-94250-5 Springer-Verlag Berlin Heidelberg New York

The IMA Volumes
in Mathematics and its Applications

Current Volumes:

Volume 1: Homogenization and Effective Moduli of Materials and Media
   Editors: Jerry Ericksen, David Kinderlehrer, Robert Kohn, J.-L. Lions

Volume 2: Oscillation Theory, Computation, and Methods of Compensated Compactness
   Editors: Constantine Dafermos, Jerry Ericksen,
   David Kinderlehrer, Marshall Slemrod

Volume 3: Metastability and Incompletely Posed Problems
   Editors: Stuart Antman, Jerry Ericksen, David Kinderlehrer, Ingo Müller

Volume 4: Dynamical Problems in Continuum Physics
   Editors: Jerry Bona, Constantine Dafermos, Jerry Ericksen,
   David Kinderlehrer

Volume 5: Theory and Applications of Liquid Crystals
   Editors: Jerry Ericksen and David Kinderlehrer

Volume 6: Amorphous Polymers and Non-Newtonian Fluids
   Editors: Constantine Dafermos, Jerry Ericksen, David Kinderlehrer

Volume 7: Random Media
   Editor: George Papanicolaou

Volume 8: Percolation Theory and Ergodic Theory of Infinite Particle Systems
   Editor: Harry Kesten

Volume 9: Hydrodynamic Behavior and Interacting Particle Systems
   Editor: George Papanicolaou

Volume 10: Stochastic Differential Systems, Stochastic Control Theory and Applications
   Editors: Wendell Fleming and Pierre-Louis Lions

Volume 11: Numerical Simulation in Oil Recovery
   Editor: Mary Fanett Wheeler

Volume 12: Computational Fluid Dynamics and Reacting Gas Flows
   Editors: Bjorn Engquist, M. Luskin, Andrew Majda

Volume 13: Numerical Algorithms for Parallel Computer Architectures
  Editor: Martin H. Schultz

Volume 14: Mathematical Aspects of Scientific Software
  Editor: J.R. Rice

Volume 15: Mathematical Frontiers in Computational Chemical Physics
  Editor: D. Truhlar

Volume 16: Mathematics in Industrial Problems
  by Avner Friedman

Volume 17: Applications of Combinatorics and Graph Theory to the Biological and Social Sciences
  Editor: Fred Roberts

Volume 18: q-Series and Partitions
  Editor: Dennis Stanton

Volume 19: Invariant Theory and Tableaux
  Editor: Dennis Stanton

Volume 20: Coding Theory and Design Theory Part I: Coding Theory
  Editor: Dijen Ray-Chaudhuri

Volume 21: Coding Theory and Design Theory Part II: Design Theory
  Editor: Dijen Ray-Chaudhuri

Volume 22: Signal Processing: Part I - Signal Processing Theory
  Editors: L. Auslander, F.A. Grünbaum, J.W. Helton, T. Kailath, P. Khargonekar and S. Mitter

Volume 23: Signal Processing: Part II - Control Theory and Applications of Signal Processing
  Editors: L. Auslander, F.A. Grünbaum, J.W. Helton, T. Kailath, P. Khargonekar and S. Mitter

Volume 24: Mathematics in Industrial Problems, Part 2
  by Avner Friedman

Volume 25: Solitons in Physics, Mathematics, and Nonlinear Optics
  Editors: Peter J. Olver and David H. Sattinger

Volume 26: Two Phase Flows and Waves
　　　　Editors: Daniel D. Joseph and David G. Schaeffer

Volume 27: Nonlinear Evolution Equations that Change Type
　　　　Editors: Barbara Lee Keyfitz and Michael Shearer

Volume 28: Computer Aided Proofs in Analysis
　　　　Editors: Kenneth Meyer and Dieter Schmidt

Volume 29: Multidimensional Hyperbolic Problems and Computations
　　　　Editors: Andrew Majda and Jim Glimm

Volume 30: Microlocal Analysis and Nonlinear Waves
　　　　Editors: Michael Beals, R. Melrose and J. Rauch

Volume 31: Mathematics in Industrial Problems, Part 3
　　　　by Avner Friedman

Volume 32: Radar and Sonar, Part 1
　　　　by Richard Blahut, Willard Miller, Jr. and Calvin Wilcox

Volume 33: Directions in Robust Statistics and Diagnostics: Part I
　　　　Editors: Werner A. Stahel and Sanford Weisberg

Volume 34: Directions in Robust Statistics and Diagnostics: Part II
　　　　Editors: Werner A. Stahel and Sanford Weisberg

Volume 35: Dynamical Issues in Combustion Theory
　　　　Editors: P. Fife, A. Liñán and F.A. Williams

Volume 36: Computing and Graphics in Statistics
　　　　Editors: Andreas Buja and Paul Tukey

Volume 37: Patterns and Dynamics in Reactive Media
　　　　Editors: Harry Swinney, Gus Aris and Don Aronson

Volume 38: Mathematics in Industrial Problems, Part 4
　　　　by Avner Friedman

Volume 39: Radar and Sonar, Part II
　　　　Editors: F. Alberto Grünbaum, Marvin Bernfeld and Richard E. Blahut

Volume 40: Nonlinear Phenomena in Atmospheric and Oceanic Sciences
  Editors: George F. Carnevale and Raymond T. Pierrehumbert

Volume 41: Chaotic Processes in the Geological Sciences
  Editor: David A. Yuen

Volume 42: Partial Differential Equations with Minimal Smoothness and Applications
  Editors: B. Dahlberg, E. Fabes, R. Fefferman, D. Jerison, C. Kenig and J. Pipher

Volume 43: On the Evolution of Phase Boundaries
  Editors: Morton E. Gurtin and Geoffrey B. McFadden

Volume 44: Twist Mappings and Their Applications
  Editor: Richard McGehee and Kenneth R. Meyer

Volume 45: New Directions in Time Series Analysis, Part I
  Editors: David Brillinger, Peter Caines, John Geweke, Emanuel Parzen, Murray Rosenblatt, and Murad S. Taqqu

Volume 46: New Directions in Time Series Analysis, Part II
  Editors: David Brillinger, Peter Caines, John Geweke, Emanuel Parzen, Murray Rosenblatt, and Murad S. Taqqu

Volume 47: Degenerate Diffusions
  Editors: W.-M. Ni, L.A. Peletier, J.-L. Vazquez

Volume 48: Linear Algebra, Markov Chains and Queueing Models
  Editors: Carl D. Meyer and Robert J. Plemmons

Volume 49: Mathematics in Industrial Problems, Part 5
  by Avner Friedman

Volume 50: Combinatorial and Graph-Theoretic Problems in Linear Algebra
  Editors: Richard Brualdi, Shmuel Friedland and Victor Klee

Volume 51: Statistical Thermodynamics and Differential Geometry of Microstructured Materials
  Editors: H. Ted Davis and Johannes C.C. Nitsche

Volume 52: Shock Induced Transitions and Phase Structures
  Editors: J.E. Dunn, Roger Fosdick and Marshall Slemrod

Volume 53: Variational and Free Boundary Problems
  Editors: Avner Friedman and Joel Spruck

Volume 54: Microstructure and Phase Transitions
  Editors: D. Kinderlehrer, R. James and M. Luskin

Volume 55: Turbulence in Fluid Flows: A Dynamical Systems Approach
  Editors: C. Foias, G.R. Sell and R. Temam

Volume 56: Graph Theory and Sparse Matrix Computation
  Editors: Alan George, John R. Gilbert and Joseph W.H. Liu

Volume 57: Mathematics in Industrial Problems, Part 6
  by Avner Friedman

Volume 58: Semiconductors, Part I
  W.M. Coughran, Jr., Julian Cole, Peter Lloyd and Jacob White

Volume 59: Semiconductors, Part II
  W.M. Coughran, Jr., Julian Cole, Peter Lloyd and Jacob White

Forthcoming Volumes:

Phase Transitions and Free Boundaries

  Free Boundaries in Viscous Flows

Applied Linear Algebra

  Linear Algebra for Signal Processing

  Linear Algebra for Control Theory

Summer Program *Environmental Studies*

  Environmental Studies

Control Theory

  Robust Control Theory

  Control Design for Advanced Engineering Systems: Complexity, Uncertainty, Information and Organization

  Control and Optimal Design of Distributed Parameter Systems

  Flow Control

  Robotics

  Nonsmooth Analysis & Geometric Methods in Deterministic Optimal Control

Systems & Control Theory for Power Systems

Adaptive Control, Filtering and Signal Processing

Discrete Event Systems, Manufacturing, Systems, and Communication Networks

Mathematical Finance

FOREWORD

This IMA Volume in Mathematics and its Applications

SEMICONDUCTORS, PART I

is based on the proceedings of the IMA summer program "Semiconductors." Our goal was to foster interaction in this interdisciplinary field which involves electrical engineers, computer scientists, semiconductor physicists and mathematicians, from both university and industry. In particular, the program was meant to encourage the participation of numerical and mathematical analysts with backgrounds in ordinary and partial differential equations, to help get them involved in the mathematical aspects of semiconductor models and circuits. We are grateful to W.M. Coughran, Jr., Julian Cole, Peter Lloyd, and Jacob White for helping Farouk Odeh organize this activity and trust that the proceedings will provide a fitting memorial to Farouk.

We also take this opportunity to thank those agencies whose financial support made the program possible: the Air Force Office of Scientific Research, the Army Research Office, the National Science Foundation, and the Office of Naval Research.

Avner Friedman

Willard Miller, Jr.

Preface to Part I

Semiconductor and integrated-circuit modeling are an important part of the high-technology "chip" industry, whose high-performance, low-cost microprocessors and high-density memory designs form the basis for supercomputers, engineering workstations, laptop computers, and other modern information appliances. There are a variety of differential equation problems that must be solved to facilitate such modeling.

During July 15–August 9, 1991, the Institute for Mathematics and its Applications at the University of Minnesota ran a special program on "Semiconductors." The four weeks were broken into three major topic areas:
1. Semiconductor technology computer-aided design and process modeling during the first week (July 15–19, 1991).
2. Semiconductor device modeling during the second and third weeks (July 22–August 2, 1991).
3. Circuit analysis during the fourth week (August 5–9, 1991).

This organization was natural since process modeling provides the geometry and impurity doping characteristics that are prerequisites for device modeling; device modeling, in turn, provides static current and transient charge characteristics needed to specify the so-called compact models employed by circuit simulators. The goal of this program was to bring together scientists and mathematicians to discuss open problems, algorithms to solve such, and to form bridges between the diverse disciplines involved.

The program was championed by *Farouk Odeh* of the IBM T. J. Watson Research Center. Sadly, Dr. Odeh met an untimely death. We have dedicated the proceedings volumes to him.

In this volume, we have combined the papers from the process modeling (week 1) and circuit simulation (week 4) portions of the program.

Processing starts with a pristine wafer of material (for example, silicon). Ion implantation and diffusion are often used to introduce conductive impurities in a controlled way. Chemical and ion etching is used to change surface features. In the case of silicon, oxide is grown when an insulator is necessary. A wide variety of techniques are used to alter, shape, remove, and add various materials that form the final complex structure. The papers on process modeling in this volume include an overview of technology computer-aided design, TCAD, as well as papers on plasma and diffusion processes used in integrated-circuit fabrication.

Circuit simulation starts with models for resistors, capacitors, inductors, diodes, and transistors as a function of input voltage or current; often the diode and transistor models are extracted from device simulations. The goal is to model accurately and rapidly the response of an aggregation of such individuals devices. Multirate integration of systems of differential-algebraic equations are an important part of large-scale circuit simulation of which "waveform relaxation" is a popular instance. The papers describe techniques for dealing with a number of circuit-analysis problems.

W. M. Coughran, Jr.
Murray Hill, New Jersey

Julian Cole
Troy, New York

Peter Lloyd
Allentown, Pennsylvania

Jacob White
Cambridge, Massachusetts

CONTENTS

Foreword xi

Preface xiii

SEMICONDUCTORS, PART I

Process Modeling

IC technology CAD overview... 1
   P. Lloyd

The Boltzmann–Poisson system in weakly collisional sheaths 17
   S. Hamaguchi, R. T. Farouki, and M. Dalvie

An interface method for semiconductor process simulation 33
   M. J. Johnson and Carl L. Gardner

Asymptotic analysis of a model for the diffusion of dopant-defect pairs 49
   J.R. King

A reaction-diffusion system modeling phosphorus diffusion 67
   Walter B. Richardson, Jr.

Atomic diffusion in $GaAs$ with controlled deviation from stoichiometry..... 79
   Ken Suto and Jun-Ichi Nishizawa

Circuit Simulation

Theory of a stochastic algorithm for capacitance extraction in
integrated circuits 107
   Yannick L. Le Coz and Ralph B. Iverson

Moment-matching approximations for linear(ized) circuit analysis 115
   Nanda Gopal, Ashok Balivada, and Lawrence T. Pillage

Spectral algorithm for simulation of integrated circuits.................... 131
   O.A. Palusinski, F. Szidarovszky, C. Marcjan, and M. Abdennadher

Convergence of waveform relaxation for RC circuits 141
   Albert E. Ruehli and Charles A. Zukowski

Switched networks 147
   J. Vlach and D. Bedrosian

SEMICONDUCTORS, PART II

Device Modeling

On the Child-Langmuir law for semiconductors
  N. Ben Abdallah and P. Degond

A critical review of the fundamental semiconductor equations
  G. Baccarani, F. Odeh, A. Gnudi and D. Ventura

Physics for device simulations and its verification by measurements
  Herbert S. Bennett and Jeremiah R. Lowney

An industrial perspective on semiconductor technology modeling
  Peter A. Blakey and Thomas E. Zirkle

Combined device-circuit simulation for advanced semiconductor devices
  *J.F. Bürgler, H. Dettmer, C. Riccobene,
  W.M. Coughran, Jr., and W. Fichtner*

Methods of the kinetic theory of gases relevant to the kinetic models for semiconductors
  Carlo Cercignani

Shock waves in the hydrodynamic model for semiconductor devices
  Carl L. Gardner

Macroscopic and microscopic approach for the simulation of short devices
  A. Gnudi, D. Ventura, G. Baccarani and F. Odeh

Derivation of the high field semiconductor equations
  P.S. Hagan, R.W. Cox and B.A. Wagner

Energy models for one-carrier transport in semiconductor devices
  Joseph W. Jerome and Chi-Wang Shu

Some Applications of asymptotic methods in semiconductor device modeling
  Leonid V. Kalachev

Discretization of three dimensional drift-diffusion equations by numerically stable finite elements
  Thomas Kerkhoven

Mathematical modeling of quantum wires in periodic heterojunction structures
  Thomas Kerkhoven

Numerical simulation of MOS transistors
  Erasmus Langer

Scattering theory of high frequency quantum transport
  H.C. Liu

Accelerating dynamic iteration methods with application to semiconductor device simulation
  Andrew Lumsdaine and Jacob K. White

Boundary value problems in semiconductors for the stationary Vlasov-Maxwell-Boltzmann equations
  F. Poupaud

On the treatment of the collision operator for hydrodynamic models
  Luis G. Reyna and Andrés Saúl

Adaptive methods for the solution of the Wigner-Poisson system
  Christian Ringhofer

The derivation of analytic device models by asymptotic methods
  Christian Schmeiser and Andreas Unterreiter

symmetric forms of energy - momentum transport models
  Michael Sever

Analysis of the Gunn effect
  H. Steinrück and P. Szmolyan

Some examples of singular perturbation problems in device modeling
  Michael J. Ward, Luis Reyna and F. Odeh

Dedication

Farouk Odeh (1933 - 1992)

Farouk Odeh died unexpectedly at Yorktown Heights, New York, on May 6, 1992, at the age of 58. He was born on July 4, 1933, at Nablus, Palestine. Odeh spent most of his distinguished career working in the Mathematical Sciences Department of the Thomas J. Watson Research Center of IBM at Yorktown Heights. He made important contributions to mathematical and numerical analysis, pure and applied mathematics, and physics.

After finishing his undergraduate education at the University of Cairo, Egypt, in 1955, Odeh lectured for a year at the Teachers College in Amman, Jordan. In 1956 he became a graduate student and Research Assistant of his thesis advisor,

B. Friedman of the Mathematics Department at the University of California at Berkeley where he received his Ph.D. in Applied Mathematics in 1960. Odeh's thesis and related work was in scattering and radiation theory. Working with J. Keller as a young post-doctoral researcher, Odeh gave the first rigorous treatment of the structure and properties of solutions to a class of partial differential equations with periodic coefficients which includes the Schroedinger equation. This early work, which was referred to often and is still being cited, provided a mathematical foundation for the theory of Bloch waves in the multidimensional case which is at the heart of crystal band theory.

Odeh joined IBM in 1960 and, with the exception of several temporary outside assignments, he remained at the Watson Research Center at Yorktown Heights, New York until his death. In 1962/63 he was a Temporary Member of the Courant Institute of Mathematical Sciences of New York University, and from 1976 through 1978 he was a Visiting Member of that institute. In 1967/68 Odeh spent a sabbatical year as a Visiting Professor at the Department of Mathematics of the American University in Beirut, Lebanon. Since 1985, Odeh was Manager of the Differential Equations Group at the Watson Research Center. During this time, several young researchers joined that group whose work now is devoted mainly to the theory and numerical solution of partial differential equations.

Mechanics, Superconductivity: During his early years at IBM, Odeh made important contributions in various areas of applied mathematics, notably in mechanics and superconductivity. Working with I. Tadjbakhsh, he studied existence problems in elasticity, and visco-elasticity. In superconductivity, he proved existence and uniqueness results for solutions to the linear differential equations of the London model, and to the integro-differential equations governing the non-local Pippard model. In cooperation with H. Cohen, Odeh studied the dynamics of the switching between the superconducting and normal conducting states. Later on, he worked on existence and bifurcation problems in the framework of the nonlinear Ginzburg-Landau theory of superconductivity and calculated magnetic field distributions and critical fields in superconducting films, based on a nonlocal generalization of the Ginzburg-Landau theory.

Ordinary Differential Equations: Starting in the 1970's, Odeh began his very fruitful work on the theory of numerical integration methods for systems of stiff ordinary differential equations, i.e. systems which simultaneously possess very fast and very slow modes. It was a great privilege for me to cooperate with him in this area of numerical analysis. Early on, Odeh gave an elegant proof of linear stability of multistep integration formulas for arbitrarily large, fixed steps by applying the powerful degree theory of analytic maps on Riemann surfaces. Around 1975, Odeh proposed a technique, based on l_2-estimates, for obtaining stability results for A-stable formulas as applied to stiff dissipative nonlinear systems of differential equations.

In solving stiff systems with smooth solutions, methods of higher accuracy and less stability are both more natural and more efficient than A-stable formulas. Such methods were in practical use for a long time but, when applied to nonlinear problems, their stability properties were not understood. In 1981, Odeh and O. Nevanlinna introduced a new analysis technique, the multiplier theory, for measuring simultaneously the reduced stability of a method and the degree of smoothness of the stiff system to which the method is applied. Integration methods are characterized by properties of their multipliers, which are special l_1-sequences associated with such methods. The multiplier determines the adequacy of a method in the nonlinear regime. By combining this theory with Liapunov and functional analysis techniques, Odeh and Nevanlinna proved convergence of methods with appropriate multipliers. They also gave strategies for selecting steps and adaptively choosing methods during the course of integration to ensure convergence of the numerical solution and to optimize efficiency.

One of the by-products of the multiplier theory is a numerical "uncertainty principle" which quantifies the incompatibility of extreme accuracy and stability requirements for multistep methods. The theory has application to the waveform relaxation method for circuit simulation discussed in the subsequent paragraph, to the discretization of hydrodynamic flows of low Reynolds numbers, and to stability aspects of multivariable control systems.

From a mathematical viewpoint, the introduction of the multiplier theory is considered, by experts in the field, as one of the most sophisticated achievements of modern numerical analysis. For this contribution Odeh received an IBM Outstanding Innovation Award in 1985.

Waveform relaxation: The advent of very large scale integrated circuits during the 1980's led to a need for efficient numerical integration methods for very large, special systems of ordinary differential equations. The waveform relaxation method, which was proposed by engineers for the computer simulation of circuits and devices, answered this need. Working with A. Ruehli, J. White and O. Nevanlinna, Odeh made important contributions to the theory and practice of the waveform relaxation method. Among the theoretical contributions was a convergence proof for the discrete version of this method. This proof made use of the multiplier theory mentioned above and was valid under physically realistic assumptions. Odeh also contributed to a deeper understanding of the waveform relaxation method and its robustness, and thereby to improvements in its practical implementation. His work on this method also led to the first A-stability theorem for multirate integration methods for ordinary differential equations of order greater than one.

Semiconductors: Odeh's work on semiconductor device simulation, which began during the 1980's, led to some of the most significant achievements of his scientific career. At the center of these are his contributions to the hydrodynamic model, a generalization of the popular drift-diffusion model which, until recently, was used successfully for semiconductor device simulation. The hydrodynamic model takes into account non-local effects, such as velocity overshoots in response to rapid variation of the electric field, and hot electron effects, such as impact ionization. These effects became important to the understanding of devices because of advances in VLSI technology and progressive device miniaturization. Device simulation rests upon the Boltzmann transport equation. The drift-diffusion model is based upon a first moment approximation to that equation. The more general hydrodynamic model on the other hand, takes into consideration the first two moments of the Boltzmann equation, yielding the charge density equation, the momentum balance equation, and the energy balance equation.

Odeh, together with M. Rudan of the University of Bologna, was the first to seriously analyze the equations of the hydrodynamic model which had been proposed during the 1970's. Odeh and his collaborator proposed the first convergent multi-dimensional discretization of the full hydrodynamic equations and moreover, this discretization could easily be incorporated in existing drift-diffusion codes. In collaboration with M. Rudan and A. Gnudi of the University of Bologna, he was also the first to simulate actual one- and two-dimensional devices using the full set of hydrodynamic equations and to achieve good agreement with experimental results. More recently, Odeh with G. Bacarrani, M. Rudan, and A. Gnudi of the University of Bologna, has proposed, analyzed, and implemented a new alternative spherical harmonics approximation to the Boltzmann equations whose range of applicability extends beyond that of the hydrodynamic model. This is a fundamental, and difficult extension which is expected to have considerable impact in the field of semiconductor device simulation.

In addition to the practical analyses and implementations mentioned briefly above, Odeh led an extensive effort to develop a mathematical foundation and theory for semiconductor design and simulation. There are several aspects to this part of his work. With M. Ward, D. Cohen, and L. Reyna, he successfully applied asymptotic techniques to obtain such a theory for the current/voltage relationships within one-dimensional and quasi-one dimensional devices. This provided both a mathematical basis for existing engineering analyses and extended that theory. With E. Thomann, he made important contributions to the question of the well-posedness of the hydrodynamic equations, determining appropriate boundary conditions for the hydrodynamic model in both the steady state and time dependent cases. With H. Steinrueck, using Wigner functions, he also made important contributions to the modelling and simulation of devices where quantum effects are important. His knowledge of both the physics and the engineering aspects of semiconductor devices, combined with his extensive mathematical knowledge and talents, made him unique within the mathematical semiconductor community.

The work of Odeh and his collaborators on the hydrodynamic model has generated an enormous literature in mathematics and in device physics and simulation. Because of his contributions, Odeh is sometimes referred to, by researchers in this field, as "the father of the hydrodynamic model". The hydrodynamic model has been incorporated in FIELDAY, the official IBM software tool for device simulation. Odeh's contributions to this field have spanned mathematical and numerical analysis, electrical engineering and quantum theory of devices and were honored by an IBM Outstanding Innovation Award in 1991.

Some of the information used in this obituary was taken from personal communications by W. Coughran, Jr., G. Dahlquist, C. Gardner, and J. Keller.

For his colleagues and many friends, the passing of Farouk Odeh is a great loss.

He was an extremely talented and creative mathematician and scientist and a very sensitive human being. All of us were enriched by our association with Farouk's warm friendship and collegiality and with his great scientific knowledge and insights. We will sorely miss him.

Werner Liniger
Thomas J. Watson Research Center

LIST OF PARTICIPANTS

Aarden, J.	University of Nijmegen
Baccarani, Giorgo	University of Bologna
Bennett, Herbert	NIST
Biswas, Rana	Iowa State University
Blakey, Peter	Motorola Corporation
Borucki, Leonard	Motorola Corporation
Buergler, Josef	ETH Zurich
Casey, Michael	University of Pittsburgh
Cercignani, Carlo	Politecnico di Milano
Cole, Dan	IBM GPD
Cole, Julian	Rensselaer Polytechnic Institute
Coughran, Jr., William	AT&T Bell Labs
Cox, Paul	Texas Instruments
Degond, Pierre	Ecole Polytechnique
Gaal, Steven	University of Minnesota
Gardner, Carl	Duke University
Gartland, Chuck	Kent State University
Gerber, Dean	IBM
Giles, Martin	University of Michigan
Glodjo, Arman	University of Manitoba
Gnudi, Antonio	Universita Degli Studi Di Bologna
Grubin, Harold	Scientific Research Associates
Hagan, Patrick	Los Alamos National Lab
Hamaguchi, Satoshi	IBM
Henderson, Mike	IBM
Jerome, Joseph W.	Northwestern University
Johnson, Michael	IBM
Kalachev, Leonid	Moscow State University
Kerkhoven, Thomas	University of Illinois, Urbana
King, John	University of Nottingham
Kundert, Ken	Cadence Design Systems
Langer, Erasmus	Technical U. Vienna
Law, Mark	University of Florida
Leimkuhler, Ben	University of Kansas
Liniger, W.	IBM
Liu, H.C.	National Research Council, Ottawa
Liu, Sally	AT&T Bell Labs
Liu, Xu-Dong	UCLA
Lloyd, Peter	AT&T Bell Labs
Lojek, Robert	Motorola
Lumsdaine, Andrew	MIT
Makohon, Richard	University of Portland

Meinerzhagen, Berndt	Technischen Hochschule Aachen
Melville, Robert	AT&T Bell Labs
O'Malley, Robert E.	Rensselaer Polytechnic Institute
Odeh, Farouk	IBM
Palusinski, O.	University of Arizona
Perline, Ron	Drexel University
Petzold, Linda R.	University of Minnesota
Pidatella, Rosa Maria	Citta' Universitaria, Italy
Pillage, Larry	University of Texas
Please, Colin	Southhampton University
Poupaud, Frederic	University of Nice
Reyna, Luis	IBM
Richardson, Walter	University of Texas at San Antonio
Ringhofer, Christian	Arizona State University
Rose, Donald J.	Duke University
Rudan, Massimo	University of Bologna
Ruehli, Albert	IBM
Schmeiser, Christian	TU-Wien-Austria
Seidman, Tom	U. of Maryland-Baltimore County
Sever, Michael	Hebrew University
Singhal, K.	AT&T Bell Labs
So, Wasin	IMA
Souissi, Kamel	IBM
Strojwas, Andre	Carnegie Mellon University
Suto, Ken	Tohoku University
Szmolyan, Peter	TU-Wien-Austria
Tang, Henry	IBM
Thomann, Enrique	Oregon State University
Venturino, Ezio	University of Iowa
Vlach, Jeri	University of Waterloo
Ward, Michael	Stanford University
White, Jacob	MIT
Wrzosek, Darek	University of Warsaw
Young, Richard A.	University of Portland

IC TECHNOLOGY CAD OVERVIEW*

P. LLOYD**

Abstract. This paper gives an overview of predictive Technology CAD tools for simulating and modeling the fabrication and electrical behavior of integrated circuits. Recent trends in the integration of process, device, and circuit simulation tools, and the current emergence of UNIX-based computing environments of networked workstations make possible user-friendly, task-based CAD systems for technology optimization, characterization, and cell design.

1. Introduction.

In the electronics industry, CAE and CAD tools/systems have played a critical role in reducing non recurring engineering (NRE) costs, improving product quality and shortening time-to-market intervals. The productivity and quality gain in semiconductor electronics industry stems from both shorter design intervals and more robust design verification keyed to process capability and has depended extensively on simulation. At the lowest level of the CAD tool hierarchy are the circuit, device and process simulators which link circuit design to fabrication. Detailed process and device simulation can play a key role in generating data for modeling of circuit performance prior to fabrication. Predictive capabilities enable early delivery of accurate compact models which are critical to high performance cell and detailed sub-system design. These Technology CAD tools for modeling the fabrication and electrical behavior of integrated circuits are rapidly gaining in maturity. Smooth interfacing and integration of the various modeling tools has been a recent trend particularly in industries where Technology CAD has become an integral part of IC development and is given organizational focus [1].

Initially, predictive Technology CAD (TCAD) tools are generally a substitute for physical experimentation to save time, effort and money, and to provide additional insight. Later, tools are integrated into a TCAD system and an optimization capability is added to aid in the evaluation of competing technology alternatives. Furthermore, it is recognized that the manufacturing process has inherent variability as do the operating and physical environments in which the products have to work. These variations cause product behavior to deviate from the nominal design resulting in a reduction of yield. Traditionally, circuits have been designed by using a worst-case approach, often sacrificing either performance or yield. Technology CAD tools can be used to make the production process and device and circuit design less sensitive to the inherent variations. Figure 1.1 illustrates various components in an TCAD system and their relationship in the context of IC manufacturing and design.

The evolution of IC technologies is primarily driven by the need for small, light, fast, low power, reliable electronic circuits in military and commercial systems and

---

*Presented 7/15/91 at the Summer Program on Semiconductors, Institute for Mathematics and Its Applications, University of Minnesota.
**AT&T Bell Laboratories, Allentown, Pennsylvania 18103.

information processing systems in particular. Currently the IC market is dominated by bipolar and MOS products. As of today in 1991, the minimum feature length in MOS ICs is 0.5-0.8 μm. It is expected to be shrunk to 0.35 μm in 1995, 0.25 μm in 1998 and 0.18 μm in 2001. While TCAD tools have successfully applied to current IC technologies, TCAD tools need to be extensively enhanced to better model the characteristics of future fine-line technologies. Device simulation tools have been supported by extensive research and development during the last two decades in the areas of physics and numerics, and the best tools have paced technology evolution. On the other hand, process simulation capabilities have followed behind cutting-edge process development due to the number of effects to be modeled, the high degree of interaction between physical mechanisms and the challenge of establishing appropriate numerical methods. Device models and circuit simulation need to retain their accuracy, robustness and speed to sustain the design of the very large circuits which become feasible as dimensions are deduced.

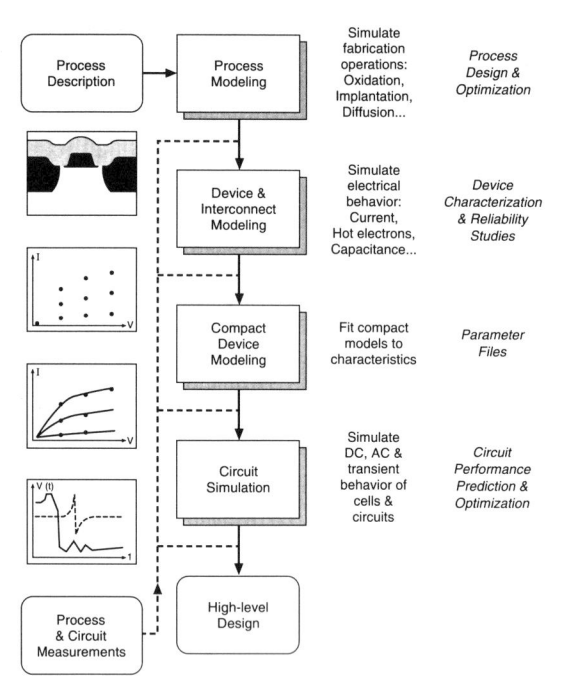

Figure 1.1. Tool Sets in Technology CAD

TCAD is maturing and several TCAD vendors have entered the market place. With support from universities, the CAD Framework Initiative (CFI) and a number of semiconductor companies, framework standards for TCAD are emerging. The advances in TCAD framework will help streamline TCAD applications. Nevertheless, there are many challenging issues yet to be addressed, for example, efficient simulation on massively parallel compute environments.

This paper provides an overview of TCAD tools and integration of these tools into TCAD systems for deployment and development. These tools include those for process, device and circuit simulation, parameter extraction and optimization. TCAD tools can be used independently or coupled together to form *tasks*. Applications of these tools and systems at AT&T to technology development and circuit design for optimization, characterization, and verification will be illustrated.

2. Process simulation.

Accurate simulation of an IC device structure begins with an accurate representation of the geometry and material properties of the structure. To obtain this representation a simulation of the individual process steps involved in fabricating the device is performed. A typical silicon IC process consists of several sequences of patterning the Si wafer, localling implanting impurities (acceptors or donors) into the exposed regions, followed by high temperature activation to get the impurities into the regions desired at the proper concentrations. Key device characteristics such as switching speed and threshold voltage depend on precise control of the doping profiles in the device. To prevent numerous iterations of processing wafers to achieve the desired device characteristics, simulation of the incorporation and diffusion of the impurities is necessary. Once the profile of acceptors and donors in the device is simulated, this profile can be used as input into a device simulator to predict device characteristics before fabrication.

Ion implantation is the most common method of locally incorporating impurities into a silicon wafer. High energy ions of the species of interest (typically boron, arsenic, or phosphorous) are accelerated and bombarded into the silicon surface. The resultant distribution of ions in the wafer is usually described by a Pearson IV distribution in the vertical direction, and an error function distribution in the lateral directions where vertical mask edges exist. The moments of the distributions are determined by the kinetic energy and the mass of the ion. Ion implantation also results in a distribution of interstitial point defects and longer range defects which must also be accounted for.

After the impurities are incorporated into the wafer, they must be activated, or moved onto lattice sites. This is done with a high temperature anneal process which not only activates the impurities, but causes diffusion of the impurities as well. The rate of diffusion is dependent on the silicon temperature, the gradient of the impurity concentration, the local electric field, and the local point defect concentration by which the impurities may move. Often the silicon surface is exposed to an oxidizing ambient during diffusion either to grow a thin oxide for the gate of the device, a thick oxide to isolate devices, or an oxide to be used as the mask layer for a future process step. This oxidation results in a movement of the surface boundary, segregation of the impurities at the interface of the growing oxide, and generation of interstitial defects which diffuse into the silicon. All of these phenomena must be taken into consideration to accurately calculate the resultant impurity profile. The diffusion equations are solved numerically on an appropriate multi-dimensional grid and with appropriate time steps so as not to add numerical errors to the solution in addition to errors introduced by assumptions on which the formulations are based.

In a typical CMOS process, n- and p-channel devices are created on the same wafer. The impurity profile of the initial p-type substrate can be used to control the threshold voltage of the n-channel device, but a deep tub of n-type impurities must be created for the p-channel device. The wafer is covered with photoresist, then the n-type tub region is exposed, implanted and diffused. Next, a new mask is placed on the surface which protects the regions where the devices will be formed, but exposes the regions between. A thick oxide to electrically isolate devices is grown at a high temperature. The mask over the active regions is stripped, and the thin gate oxide is grown. This is protected immediately with a layer of polysilicon for the gate material which serves as the mask in the next step for implanting a high concentration of acceptors into regions on either side of the p-channel gate to create the source and drain regions of the device. The same is then done for the n-channel device with a high concentration of donor ions. The wafer is annealed to activate these source and drain regions, after which the impurity profiles in the devices are essentially formed. Once simulated in a process simulator [2], as shown in Figure 2.1 these profiles are stored in disk files and can be used by a device simulator to calculate current-voltage and charge-voltage characteristics.

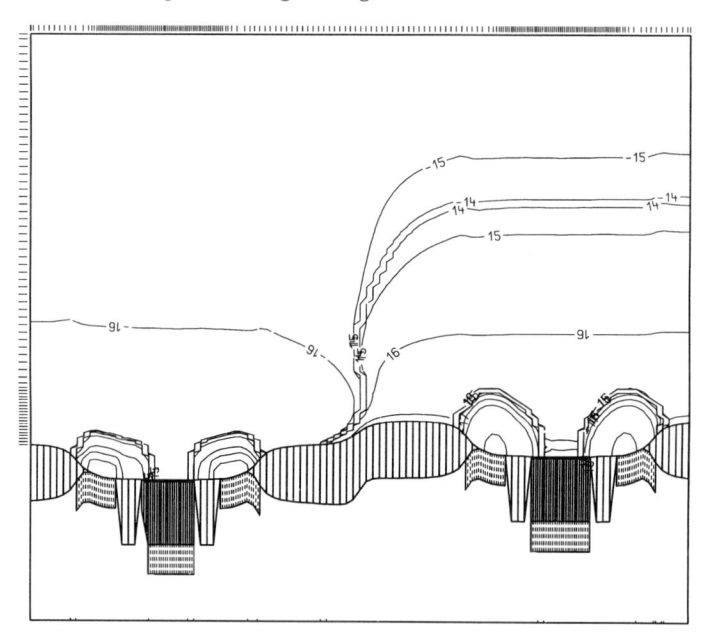

Figure 2.1. Simulated Profiles of a CMOS Device Structure by BICEPS

As can be seen from the figure, lateral dimensions of the gate and spacer regions of the devices have shrunk below $1\mu m$, and vertical dimensions of source and drain junctions are less than $.5\mu m$. Therefore, the physical details of the transient diffusion that occurs in the very short time scales which are used to achieve such dimensions are increasingly important in modeling the submicron devices. Also, the exact structure of masking and oxide layers, as a result of deposition and etching processes above the silicon, plays a significant role in determining the lateral profile of the devices. Simulation of these processes is becoming essential. Additional, dimensions are shrinking in all three dimensions, requiring three-dimensional simulation with appropriate visualization.

3. Device modeling.

As semiconductor devices continue to shrink in size, and as new technologies energy, device structures become more complicated and the need for physically-based numerical device simulation grows. Simulation tools are needed for the design of devices and in order to gain insight into new physical effects. At present, device modeling has become a necessary and integral element in any new process or technology development effort.

Device modeling is accomplished by solving the basic equations governing the behavior of semiconductor devices. Basically, Maxwell's equations of electromagnetism and the Boltzmann Transport Equation (BTE).

A direct approach to solve BTE is the Monte Carlo method. This technique simulates, at a microscopic level, the transport process of mobile carriers. The Monte Carlo approach has proven to be successful in simulating transport effects. However, its primary drawback is the enormous cost associated with the long cpu time required, particularly when coupled with Poisson's equation. The Hydrodynamic model or the Momentum and Energy Balance equations are alternative approaches to solving the BTE. However, the simplest form of the transport equation is that of the Drift and Diffusion model. This model, which can be derived from the hydrodynamic model, comprises electron and hole current continuity equations coupled with Poisson's equation.

The inputs to device simulation tools are typically a description of impurity doping profiles obtained from process simulation, as discussed in the previous section, and device geometry as well as bias conditions. The output will be the electrical responses e.g. steady-state, transient or small-signal waveforms of currents and voltages at terminals and/or carrier density, electric field or potential distributions inside the device.

General two-dimensional (2D) or three-dimensional (3D) device simulation programs, such as PADRE [3], as well as application-specific tools, such as MEDUSA [4], can be used to solve a wide range or problems.

There are many examples of device simulations applied in both device optimization and reliability improvement. In device optimization, various devices are simulated to quantify the effects of short channel length and narrow channel widths using 2D and 3D device simulators. As far as reliability improvement is concerned, device simulations have been used to refine CMOS device structure to prevent latchup problems [3] and gain insight into the effects of hot carriers and velocity overshoot.

As dimensions of electronic devices decrease, resulting in faster switching speed, delay caused by parasitic capacitance and resistance of interconnections becomes more significant. The RESCAL program [5] has been developed to solve Laplace's equation in two dimensions to provide fast and accurate values of distributed capacitance and resistance. RESCAL also produces plots of equipotential and flux lines to represent visually the distribution of the electric field. Figure 3.1 shows an example of a contour plot for a structure in which there are two layers of different dielectric constants, and two trapezoid-shaped conductors in the lower dielectric layer over a conducting substrate.

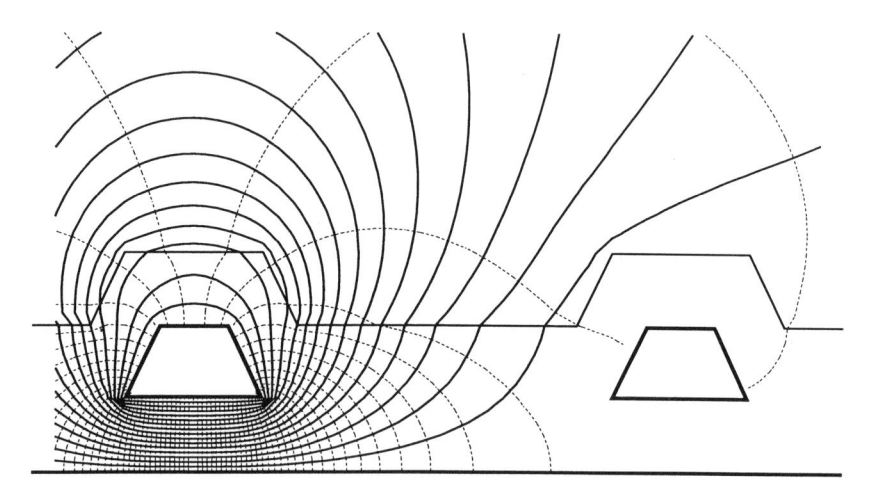

Figure 3.1. Contour Plot of the Electrostatic Field for two Conductors over a Ground Plane and with two Dielectrics

4. Compact models.

Circuit simulations are used to verify IC designs based upon compact device models, so the models must be able to accurately represent characteristics of devices being manufactured. Also, the device characteristics generated from compact models provide a reference to which the manufacturing process should be controlled such that the device characteristics will resemble the reference. Compact models are thus an important link between IC design and manufacturing.

Aggressive IC design places strong demands on compact device models. The models must allow to accurately represent the DC, AC and transient behavior of circuits, and often also the circuit noise performance, distortion level and sensitivity to variations in manufacturing and operating conditions. In addition, compact models must be computationally efficient, must be simple enough for robust parameter extraction, and must model device behavior over a wide range of bias, geometry and operating temperature. To meet all these challenges is often difficult, for example, MOSFETs exhibit an exponential variation of current with applied bias in the subthreshold region of operation and a polynomial variation of current with applied bias above threshold. It is also desirable for compact models to have a good basis in device physics, so that physical understanding can be used to guide model development and ongoing model improvements, and so that process variations measured as changes in test wafer measurements can be mapped, at least to a first order, into changes in model parameters.

ASIM3, an enhanced version of the model described in [6], is the most advanced MOSFET model available in the AT&T circuit simulator, ADVICE [7]. ASIM includes subthreshold conduction, models short and narrow channel effects, and is based on an advanced mobility model that accounts for mobility reduction due to gate and backgate fields and due to velocity saturation. ASIM is charge-based, and accurately models both the partitioning of charge between the source and drain and the variation of overlap capacitance with bias. Both the geometry and temperature dependence of MOSFET behavior are modeled by ASIM. ASIM includes models for noise and substrate injection current. The current and charge models of ASIM are continuous in function value and derivatives with respect to the applied biases across all operating regions.

Figure 4.1 shows the output and subthreshold characteristics of ASIM, compared with data from the MEDUSA device simulator. ASIM accurately models the DC current, output conductance and transconductance of MOSFETs.

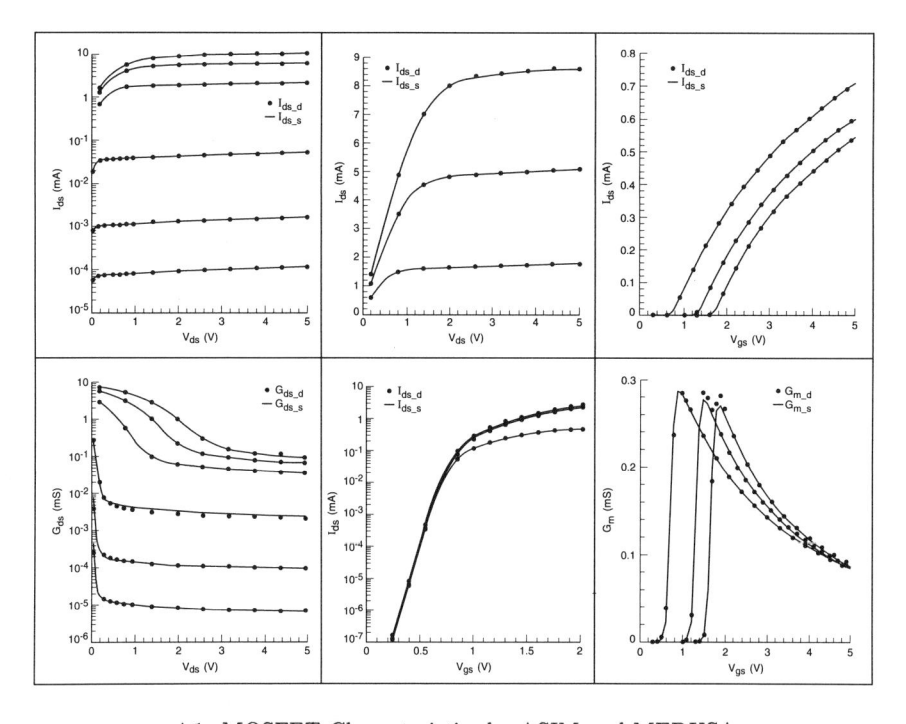

4.1. MOSFET Characteristics by ASIM and MEDUSA

Major deficiencies have existed in previous MOSFET models. First, most MOSFET models are formulated as regional models, that is, different modeling equations are used in the subthreshold, triode and saturation regions of operation. Regional models have limited continuity, and display kinks and glitches, at the region boundaries. This causes problems for parameter extraction and DC convergence, limits the accuracy of distortion analyses, makes some advanced techniques such as homotopy [8] inapplicable to MOS circuits, limits the order of integration that can be used for transient analyses, and leads to inefficient transient analyses as it causes small time-steps to be used. Second, most MOSFET models are formulated with the source node as the reference. This easily causes the model to display asymmetries with respect to the source and drain, even though MOSFETs usually are symmetric devices.

5. Circuit simulation.

Small feature size and mixed analog and digital components in today's VLSI IC technologies demand more accurate technology modeling in circuit simulation than that in the past. The emergence of networked workstation environments demands, flexibility in task-oriented procedural simulation and robustness in design centering.

A circuit simulation system normally consists of six components: front-end for user interactions, macro interface for procedural threads, design centering and optimization, analysis engine, model interface for device models, and graphics display.

- The Front-End

    The front-end provides a graphic user interface between the user and the circuit simulator. In addition, it interacts with a schematic capture program and also provides remote execution capabilities across the network. User-friendliness is determined by not only the set of commands but also its look and feel.

- The Macro Interface

    Through this interface, designers may drive the simulator by programming a sequence of circuit simulation tasks, including updating the circuit, performing simulation and analyzing results, thus eliminating repetitive manual procedures. AT&T's circuit simulator ADVICE contains a C-language based macro interface an its procedural simulation controls the circuit simulation at fine granularity at the simulation-command level by linking the engine and the interface through UNIX pipes. ADVICE also has a specification-driven generator [9] which generates C code for macro procedural simulation, thus it relieves designers from the burden of mastering programming tasks.

- The Design Optimization

    Successful circuit design often requires design iterations, to optimize the circuit. Normally the optimizer is coupled closely with the macro interface to control the analysis engine. The simulation language is extended to allow specifications of design objectives. Statistical models [10] for semiconductor devices are needed for yield and manufacturability analysis to determine the impact of manufacturing and environmental variations on a design. Various optimization algorithms can be used. For example, ADVICE contains interactive features for design optimization and manufacturability analysis based on a software system called CENTER [11]. The CENTER system contains features of both deterministic optimization and manufacturability analysis. The deterministic optimization in CENTER can be accomplished by either sequential quadratic programming or random search.

- The Analysis Engine
  The engine provides analysis capabilities for dc, ac, noise, sensitivity, transient, steady-state, Monte-Carlo, loop stability, and others. It handles a comprehensive set of components including resistors, capacitors, inductors, transformers, voltage sources, current sources, controlled sources, diodes, bipolar transistors (bjts), junction field-effect transistors (jfets), mosfets, josephson junction devices, switches, and transmission lines. For most types of components, there is an associated model with a set of model parameters to characterize a specific IC technology. For example, ADVICE contains an extended Gummel-Poon model [12] and a charge-based short channel model, ASIM which accurately model the devices in AT&T's technologies. Solving the circuit equations involves a number of numerical techniques, such as implicit time integration, Newton-Raphson iterations, homotopy methods, and sparse matrix techniques.

- The Model Interface
  The model interface allows users to introduce their own models into the simulator [13]. It has proven to be an effective tool for developing new models for circuit simulation.

- The Graphics Display
  Graphics display is as important as the front-end. It allows the user to visually examine the simulation results mostly via 2-D graphics.

6. TCAD system framework.

In the area of TCAD frameworks, a key issue is the distinction between frameworks used primarily for (1) tool development and (2) tool integration/deployment. In order to better differentiate these two, we will refer to a tool development *environment* and a tool deployment *framework*.

In recent years, TCAD tools have matured to the extend that they are capable, with moderate accuracy, of modeling the dominant performance-limiting features of current technologies. With this maturity has come an increasing demand for accuracy, robustness, integration (with other tools) and ease of use. These demands place additional constraints on TCAD tool developers and require an increase in development resources.

This situation has been recognized by several university development groups, as well as their industrial counterparts. One of the early efforts in this area was the *Profile Interchange Format* (PIF) [14] which is now becoming a standard for data exchange between TCAD simulation tools. Data exchange between TCAD tools is the first hurdle that must be overcome before tools can be integrated. Integration, however, is only the first stage of deployment. Other issues such as capability, accuracy, robustness, ease of use, and user-friendliness all play an important role in gaining the user's acceptance.

TCAD frameworks can be viewed from two distinct points of view:

- As a framework for the *integration* of TCAD tools. An example of such a framework is the AT&T Mecca system [15] and Intel's Ease system [16] which integrate process/device and circuit simulation with analysis tools such as optimization and parameter extraction as shown in Figure 6.1.
- As an environment, and associated set of support tools, for the *development* of TCAD tools as shown in Figure 6.1. We are not aware of any significant previous work in this area.

The integration framework is of immediate benefit to all TCAD practitioners, as well as TCAD customer organizations: technology development, technology characterization, and manufacturing. The development environment, on the other hand, is of importance primarily to universities and industry R&D groups engaged in TCAD tool development.

The next two subsections expand on the above definitions.

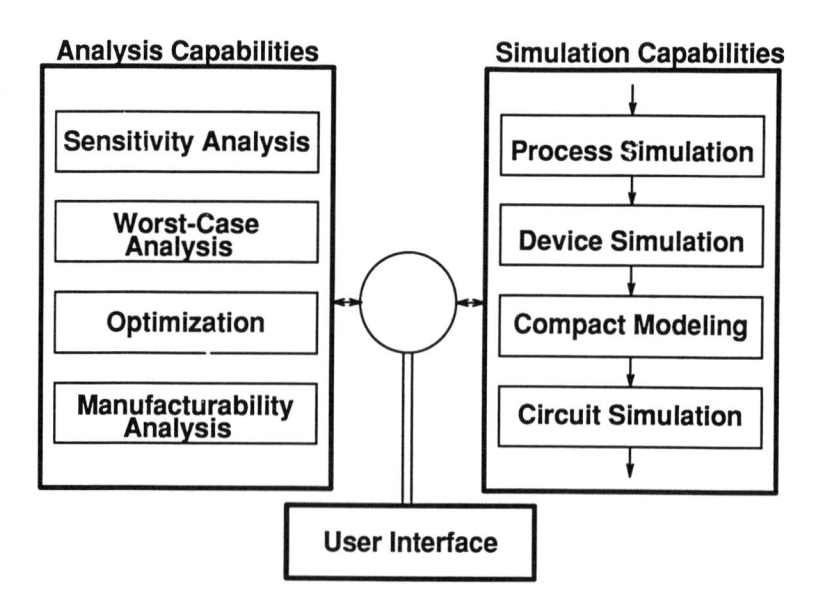

Figure 6.1. An Integrated set of TCAD Simulation and Analysis Tools

6.1 Summary. We have discussed two views of TCAD frameworks. These two views are not exclusive. In fact, achieving a standard for TCAD tool development would result in increased uniformity across the tools, which would greatly ease the integration of such tools into a common TCAD system.

In addition to the traditional uses of TCAD in technology development and characterization, opportunities are being persued in:

Process Control and Diagnosis

> Where TCAD tools can be used to develop algorithms for active process control, and as an aid in diagnosing process faults.

Computer Integrated Manufacturing

> Where TCAD tools can be integrated into the manufacturing environment and used to evaluate and enhance reliability, yield and manufacturability.

7. Applications (Loop-closure, Optimization, Worst case). Figure 1.1 illustrates how the individual AT&T technology CAD tools are put together into an integrated system. With this system, given the process description and the structure and geometry of a device, a compact device model parameters can be determined. The compact model can be used in the ADVICE circuit simulator to characterize the circuit performance. Furthermore, the effects of statistical variations in the process control parameters on the compact model parameters and the circuit performance can be determined. (Important process control parameters include furnace times and temperatures, ion implant doses and energies, etc.) This technology CAD system is a useful aid in predicting performance of circuits, verifying designs, and developing new or modified technologies.

Extraction of compact circuit model parameters is routinely done for each technology and technology variant. The extraction is carried out for at least three cases: nominal, worst-case slow and worst-case fast. For nominal, all the process input conditions are nominal. For worst-case fast, the process input conditions are such as to result in the slowest possible circuit performance (for digital MOS circuits, this amounts to minimum current drive of the transistors), while for worst-case fast, the process inputs are such as to result in the fastest possible circuit performance (maximum current drive of the transistors). Thus, each technology is characterized by a nominal, worst-case slow, and a worst-case fast compact device model. These models are used in circuit simulators to characterize important or characteristic subcircuit modules, and the worst-case models are used to verify that designs will meet their specification limits. Finally, the accuracy of the models is verified versus test data from the manufacturing line.

Optimization can be done using the CENTER software system, which exploits features of the UNIX operating system. CENTER can be used to optimize IC designs and semiconductor device technologies. But only the latter will be discussed here. Figure 7.1 shows how CENTER is integrated into AT&T's technology CAD system. The inputs include the technology objectives and constraints. CENTER contains six numerical methods for nonlinear optimization:

- Sequential quadratic programming
- A projected, augmented Lagrangian algorithm
- A quasi-Newton solver
- A nonlinear least squares minimizer
- The Nelder-Mead simplex method
- Simulated annealing

Verification of the compact device models is done based upon measurements on each lot in the manufacturing line. Figure 7.2 shows plots for the $0.9 \mu m$ CMOS technology of the measured NMOS I_{on} for 23 lots versus the I_{on} predicted by the nominal and the worst-case slow and fast compact models. The measured data are within the limits predicted by the models. Figure 7.3 shows the probability distribution of the measured ring oscillator frequency for the $1.25 \mu m$ CMOS technology versus the frequencies predicted by the compact circuit models. The predicted and measured frequencies agree well.

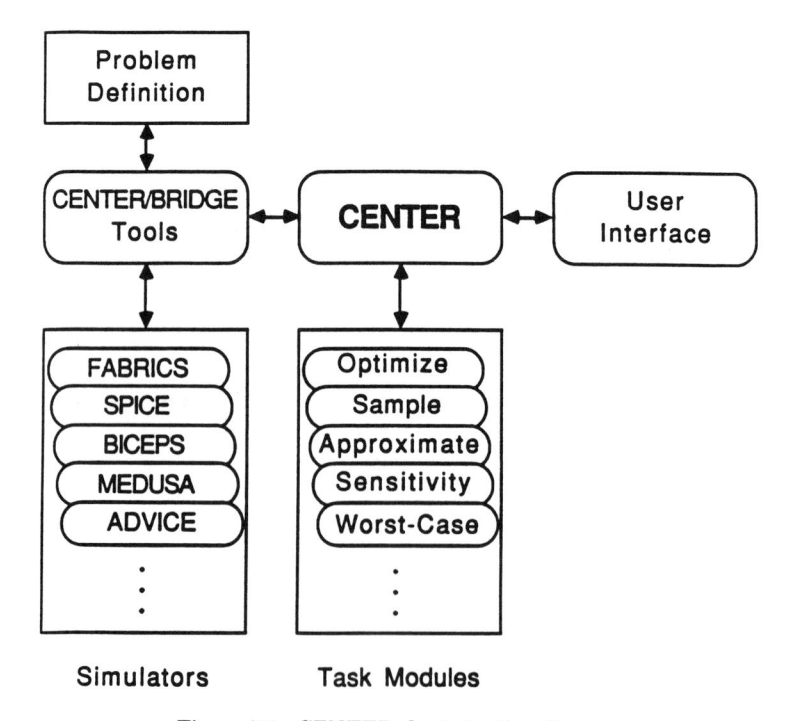

Figure 7.1. CENTER Optimization System

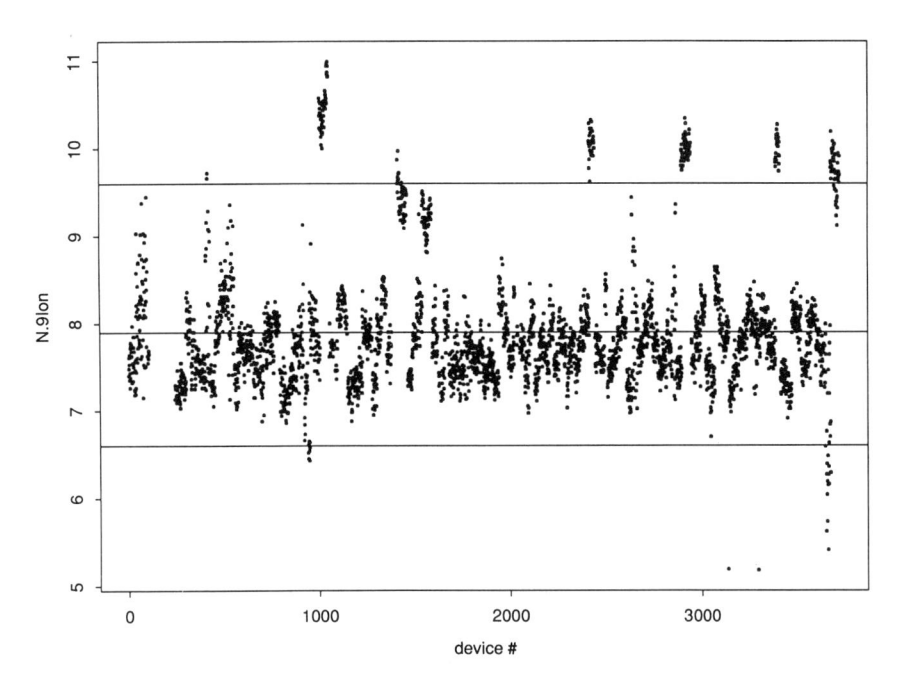

Figure 7.2. Predicted and Measured I_{on} for $0.9\mu m$ N-Channel Devices

Ring Oscillator Frequencies
1.25 µm CMOS Technology

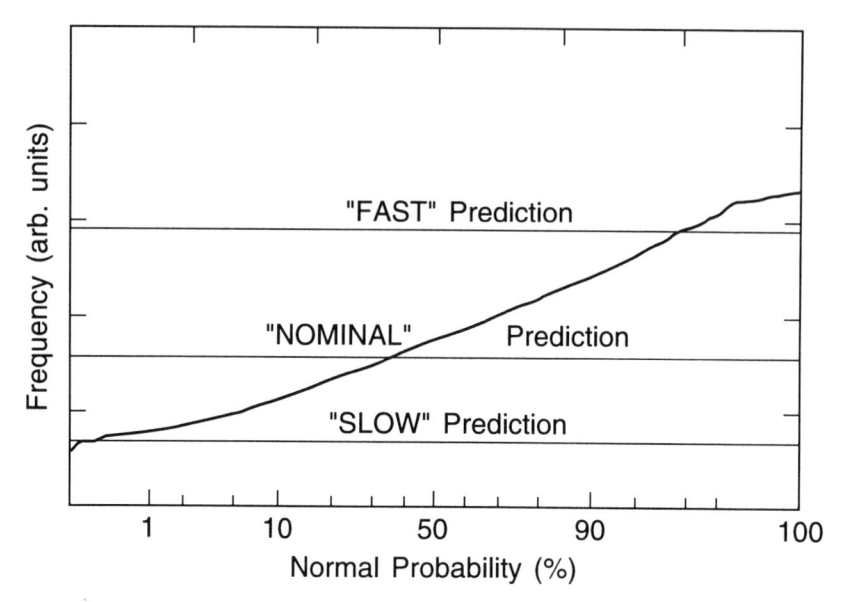

Figure 7.3. Predicted and Measured Ring Oscillator Frequencies

8. Current trends and conclusions.

The following summarizes the major observations and trends discussed in this paper:

- TCAD is maturing. Several vendors have entered the market place.
- Process simulation tools require better physical models and numerical methods.
- Above-silicon simulation and visualization are increasingly important.
- Device and interconnect models are being improved to meet challenges for submicron design.
- Circuit simulation will become more robust and efficient for large circuits with extensive interconnects.
- Algorithms are needed to exploit evolving vector parallel computer architecture.
- Evolving standards and integration framework provide for collaborations amongst companies and universities..
- Technology CAD will contribute to manufacturing studies.

9. Acknowledgements.

I thank past and present members of the Technology CAD Department at AT&T Bell Laboratories for their contributions to the work described here. In addition, I thank Heinz Dirks, Sally Liu, Jim Prendergast and Kishore Singhal for their help in preparing this paper.

REFERENCES

[1] P. LLOYD, H.K. DIRKS, E.J. PRENDERGAST, AND K. SINGHAL, *Technology CAD for Competitive Products*, IEEE Trans. Computer-Aided Design, vol. CAD-9, Nov. 1990.

[2] B.R. PENUMALLI, *A Comprehensive Two-Dimensional VLSI Process Simulation Program, BICEPS*, IEEE Trans. Electron Devices, vol. ED-30, Sept. 1983.

[3] M.R. PINTO, W.M. COUGHRAN, JR., C.S. RAFFERTY, AND E. SANGIORGI, *Device Simulation for Silicon ULSI*, Computational Electronics, Ed. K. Hess, J.P. Leburton, and U. Ravaioli, Kluwer Academic Publishers, 1991.

[4] W.L. ENGL, R. LAUR, AND H.K. DIRKS, *MEDUSA–A Simulator for Modular Circuits*, IEEE Trans. Computer-Aided Design, vol. CAD-1, April 1982.

[5] B.R. CHAWLA AND H.K. GUMMEL, *A Boundary Technique for Calculation of Distributed Resistance*, IEEE Trans. Electron Devices, vol. ED-17, Oct. 1970.

[6] S.W. LEE AND R.C. RENNICK, *A Compact IGFET Model–ASIM*, IEEE Trans. Computer-Aided Design, vol. CAD-7, Sept. 1988.

[7] L.W. NAGEL, *ADVICE for Circuit Simulation*, Proc. ISCAS, Houston, 1980.

[8] L. TRAJKOVIC, R.C. MELVILLE, S.-C. FANG, *Improving DC Convergence in a Circuit Simulator Using a Homotopy Method*, IEEE Custom Integrated Circuits conference – CICC-91, San Diego, CA, May 1991.

[9] M.S. TOTH, *MakCal: An Application Generator for ADVICE*, AT&T Technical Journal, vol. 70, Jan./Feb. 1991.

[10] S. LIU AND K. SINGHAL, *A Statistical Model for MOSFETS*, IEEE International Conference on Computer-Aided Design – ICCAD-85, Santa Clara, CA, Nov. 1985.

[11] K. SINGHAL, C.C. MCANDREW, S.R. NASSIF, AND V. VISVANATHAN, *The CENTER Design Optimization System*, AT&T Technical Journal, vol. 68, May/June 1989.

[12] G.M. KULL, L.W. NAGEL, S.-W. LEE, P. LLOYD, E.J. PRENDERGAST, AND H.K. DIRKS, *A Unified Circuit Model for Bipolar Transistors including Quasi-Saturation Effects*, IEEE Trans. Electron Devices, vol. ED-32, June 1985.

[13] S. LIU, K.C. HSU, AND P. SUBRAMANIAM, *ADMIT-ADVICE Modeling Interface Tool*, Proc. 1988 Custom Integrated Circuits Conference, Rochester, 1988.

[14] S.G. DUVALL, *An Interchange Format for Process and Device Simulation*, IEEE Trans. Computer-Aided Design, vol. CAD-7, July 1988.

[15] E.J. PRENDERGAST, *An Integrated Approach to Modeling*, Proc. NASCODE IV, Dublin, June 1985.

[16] J. MAR, K. BHARGAVAN, S.G. DUVALL, R. FIRESTONE, D.J. LUCEY, S.N. NANGAONKAR, S. WU, K.-S. YU, AND F. ZARBAKHSH, *EASE–An Application-Based CAD System for Process Design*, IEEE Trans. Computer-Aided Design, vol. CAD-6, Nov. 1987.

THE BOLTZMANN–POISSON SYSTEM
IN WEAKLY COLLISIONAL SHEATHS

S. HAMAGUCHI*, R. T. FAROUKI*, AND M. DALVIE*

Abstract. Ion distribution functions in weakly-collisional direct-current (DC) sheaths and collisionless radio-frequency (RF) sheaths are discussed from the viewpoint of kinetic theory. Analytical formulae for the ion distributions in a self-consistent field are obtained for weakly-collisional sheaths, and are shown to be in good agreement with results from Monte Carlo simulations. A detailed knowledge of the angular and energy distribution of the ion flux impinging on a surface is of interest in the plasma-processing of semiconductor materials.

1. Introduction. In various plasma processes used in integrated-circuit (IC) fabrication technology [1] [2], the outcome of an etch or deposition step can be sensitively dependent on the angular and energy distribution of the ion flux bombarding the semiconductor wafer. As the dimensions of ICs diminish and more stringent control over feature shapes and sizes is required, weakly-collisional or collisionless plasma discharges are increasingly used for such processes, since the uni-directionality of the ion flux increases as the frequency of ion-neutral collisions in the sheath diminishes. In such weakly-collisional/collisionless plasmas, however, the incident-ion distributions are expected to be sensitive functions of such basic plasma parameters as the applied cathode voltage, sheath thickness, and bulk plasma temperature.

The goal of this paper is to determine the relation between these controllable plasma parameters and the incident-ion distribution functions. To achieve this, we discuss ion kinetics in a weakly-collisional direct-current (DC) sheath, based on a boundary value problem of the steady-state Boltzmann–Poisson system. Assuming that elastic hard-sphere collisions are the dominant form of ion-neutral encounter and that the ion mean-free-path is large compared to the sheath thickness, we obtain the ion velocity distribution function at the cathode to lowest order in the collisionality parameter (i.e., the ratio of the sheath thickness to the ion mean free path). Subsequently, angular and energy distributions of the ion flux are calculated and compared with data from Monte Carlo simulations. We also briefly discuss a time-periodic Vlasov–Poisson system that describes the ion kinetics in a collisionless radio-frequency (RF) sheath. In the regime of high RF frequency, we obtain a simple expression for the characteristic "double-peaked" profile of the ion energy distribution [3] [4].

This paper is organized as follows: in the following section, the Boltzmann–Poisson system for a collisional DC sheath is formulated and solved in the limit of small collisionality. In sections 3 and 4, the angular and energy distributions of the incident-ion flux in a DC sheath are calculated analytically and numerically. In section 5, collisionless RF sheaths are discussed briefly. The final section contains conclusions.

2. Kinetic equations for DC sheaths. As a model of a DC plasma sheath [5][6], we consider a steady-state ($\partial/\partial t = 0$), capacitively coupled planar discharge

---
* IBM Thomas J. Watson Research Center, Yorktown Heights, NY 10598

with the electric field in the z direction. For the sake of brevity, the sheath is assumed to be composed purely of ions; we define the presheath/sheath boundary as the point beyond which electrons are significantly depleted. Since, in most planar discharges, the neutral density n_g is much higher than the ion density n, we take into account only two-body ion–neutral collisions. We also assume that the ions and the neutrals have equal mass, and that the neutrals are uniformly distributed and cold, i.e., the velocity distribution of the neutrals is given by $F(\mathbf{v}) = n_g \delta(\mathbf{v})$.

The ion distribution function $f(\mathbf{v}, z)$ is then governed by the following non-dimensionalized Boltzmann–Poisson system:

$$u_z \frac{\partial \bar{f}}{\partial \zeta} - \frac{d\phi}{d\zeta}\frac{\partial \bar{f}}{\partial u_z} = \frac{d}{\lambda_{mfp}} u \left\{ \int \left(\frac{u'}{u}\right)^4 \bar{f}(\mathbf{u}', \zeta) \frac{\sigma}{\sigma_{total}} d^2\Omega - \bar{f}(\mathbf{u}, \zeta) \right\}, \tag{1}$$

$$\frac{d^2\phi}{d\zeta^2} = -\int \bar{f}(\mathbf{u}, \zeta) d^3\mathbf{u}, \tag{2}$$

where $u = |\mathbf{u}|$ and $u' = |\mathbf{u}'|$. Here we have used the normalizations

$$\begin{aligned} f &= \bar{f} n_I / (\omega_{pi} d)^3, & \Phi &= \phi q n_I d^2 / \varepsilon_0, \\ \mathbf{v} &= \mathbf{u} \omega_{pi} d, & z &= \zeta d, \end{aligned} \tag{3}$$

where n_I is the ion density at the presheath/sheath boundary, $n_I = n(0)$, d is the sheath thickness, q is the ion charge, $\omega_{pi} = (q^2 n_I / \varepsilon_0 m)^{1/2}$ is the ion plasma frequency, z is distance into the sheath measured from the presheath/sheath boundary, and Φ is the electric field potential. In Eq. (1), σ is the differential cross section for ion-neutral collisions, σ_{total} is the total cross section, and $\lambda_{mfp} = (n_g \sigma_{total})^{-1}$ is the mean free path. The primed quantity $\mathbf{u}'$ denotes an ion velocity before a collision. Here the ion–neutral collisions are assumed to be elastic hard-sphere collisions. Then the differential cross section for scattering of ions into the solid angle element $d^2\Omega$ may be written [7] as

$$\sigma d^2\Omega = \frac{\sigma_{total}}{8\pi} \sin\frac{\chi}{2} d\chi d\psi, \tag{4}$$

where χ is the polar scattering angle measured in the laboratory system, and ψ is the azimuthal scattering angle about the $\mathbf{u}'$ direction. Note that $u = u' \cos \chi$.

We now solve Eqs. (1) and (2) in the limit of weak collisionality, $\varepsilon = d/\lambda_{mfp} \ll 1$. For the sake of simplicity, the ions are assumed to enter the sheath with a fixed velocity v_B as a beam, so the boundary conditions for f are given by

$$\begin{aligned} \bar{f}(\mathbf{u}, \zeta = 0) &= \frac{\delta(u_\perp)}{2\pi u_\perp} \delta(u_z - u_B) & \text{for } u_z \geq 0, \\ \bar{f}(\mathbf{u}, \zeta = 1) &= 0 & \text{for } u_z < 0. \end{aligned} \tag{5}$$

Here $u_B = v_B/\omega_{pi} d > 0$, and $u_\perp$ is the magnitude of the component of $\mathbf{u}$ perpendicular to the z direction. It is known that the initial ion stream velocity v_B is typically given by the ion sound speed $v_B = (k_B T_e/m)^{\frac{1}{2}}$, where k_B is the Boltzmann constant, T_e is the electron temperature of the bulk plasma and m is the ion mass (this is the

Bohm sheath criterion; see [8]). The boundary values for the potential are given by $\phi = 0$ and $d\phi/d\zeta = \bar{E}_I$ at $\zeta = 0$, where $\bar{E}_I$ denotes the (normalized) magnitude of the electric field at the presheath/sheath boundary.

Assuming that the dependence of $\bar{f}$ and ϕ on ε is analytic, we expand the ion distribution function $\bar{f}$ and the potential ϕ in terms of the small parameter ε in the form $\bar{f} = \bar{f}_0 + \varepsilon \bar{f}_1 + \cdots$ and $\phi = \phi_0 + \varepsilon \phi_1 + \cdots$. To the lowest order, we obtain from Eqs. (1) and (2) the following equations for a collisionless sheath:

$$u_z \frac{\partial \bar{f}_0}{\partial \zeta} - \frac{d\phi_0}{d\zeta} \frac{\partial \bar{f}_0}{\partial u_z} = 0, \tag{6}$$

$$\frac{d^2 \phi_0}{d\zeta^2} = -\int \bar{f}_0(\mathbf{u}, \zeta) d^3\mathbf{u}. \tag{7}$$

With the use of the new independent variable

$$\mathcal{E} = \frac{1}{2} u_z^2 + \phi_0(\zeta), \tag{8}$$

Eq. (6) becomes $u_z \partial \bar{f}_0(\mathbf{u}_\perp, \mathcal{E}, \zeta)/\partial \zeta = 0$ and its solution with the boundary condition (5) is given by

$$\bar{f}_0 = \begin{cases} \frac{\delta(\mathbf{u}_\perp)}{2\pi u_\perp} \delta(\sqrt{2\mathcal{E}} - u_B) & \text{if } u_z > 0 \text{ and } \mathcal{E} \geq 0, \\ 0 & \text{otherwise}. \end{cases} \tag{9}$$

The lowest–order potential ϕ_0 may be calculated through substitution of Eq. (9) into Eq. (7), i.e.,

$$\frac{d^2 \phi_0}{d\zeta^2} = \frac{-u_B}{\sqrt{u_B^2 - 2\phi_0}}. \tag{10}$$

The exact, closed–form solution of Eq. (10) is derived in [10], where it is shown that ϕ_0 is a non-positive ($\phi_0 \leq 0$) monotonically decreasing function for all $\zeta \geq 0$. It is known that in the limit $u_B \ll -2\phi_0$, the solution of Eq. (10) gives the collisionless Child–Langmuir law [9] [10]:

$$\phi_0 \propto \zeta^{4/3}.$$

We now proceed to the first-order equations, which are given by

$$u_z \frac{\partial \bar{f}_1}{\partial \zeta} - \frac{d\phi_0}{d\zeta} \frac{\partial \bar{f}_1}{\partial u_z} - \frac{d\phi_1}{d\zeta} \frac{\partial \bar{f}_0}{\partial u_z} = B^+ - B^-, \tag{11}$$

$$\frac{d^2 \phi_1}{d\zeta^2} = -\int \bar{f}_1(\mathbf{u}, \zeta) d^3\mathbf{u}, \tag{12}$$

where

(13)
$$B^+ = u \int \left(\frac{u'}{u}\right)^4 \bar{f}_0(\mathbf{u}') \frac{\sigma}{\sigma_{total}} d^2\Omega$$
$$= \begin{cases} \frac{1}{\pi u_z} \delta(\sqrt{h} - u_B) & \text{if } u_z > 0 \text{ and } h \equiv \frac{(u_\perp^2 + u_z^2)^2}{u_z^2} + 2\phi_0 \geq 0, \\ 0 & \text{otherwise.} \end{cases}$$

and

(14)
$$B^- = \begin{cases} u \frac{\delta(u_\perp)}{2\pi u_\perp} \delta(\sqrt{2\mathcal{E}} - u_B) & \text{if } u_z > 0 \text{ and } \mathcal{E} \geq 0, \\ 0 & \text{otherwise.} \end{cases}$$

For the derivation of Eq. (13), see [6].

In order to solve the first-order equation (11), we again transform the variable u_z to $\mathcal{E}$ defined in Eq. (8). It is convenient to split $\bar{f}_1$ as $\bar{f}_1 = \bar{f}_1^+ + \bar{f}_1^-$ in such a way that Eq. (11) may be written as $u_z \partial \bar{f}_1^+ / \partial \zeta = B^+$ and $u_z \partial \bar{f}_1^- / \partial \zeta - (d\phi_1/d\zeta) \partial \bar{f}_0 / \partial u_z = -B^-$, or

(15)
$$\frac{\partial \bar{f}_1^+}{\partial \zeta} = \begin{cases} \frac{1}{\pi u_z^2} \delta(\sqrt{h} - u_B) & \text{if } u_z > 0 \text{ and } h \geq 0, \\ 0 & \text{otherwise,} \end{cases}$$

with $u_z^2(\mathcal{E}, \zeta) = 2(\mathcal{E} - \phi_0(\zeta))$, and

(16)
$$\frac{\partial \bar{f}_1^-}{\partial \zeta} - \frac{1}{u_z} \frac{d\phi_1}{d\zeta} \frac{\partial \bar{f}_0}{\partial u_z} = \begin{cases} -\frac{\delta(u_\perp)}{2\pi u_\perp} \delta(\sqrt{2\mathcal{E}} - u_B) & \text{if } u_z > 0 \text{ and } \mathcal{E} \geq 0, \\ 0 & \text{otherwise,} \end{cases}$$

where $u\delta(u_\perp) = u_z \delta(u_\perp)$ is used. Since

(17)
$$-\frac{1}{u_z} \frac{d\phi_1}{d\zeta} \frac{\partial \bar{f}_0}{\partial u_z} = \frac{d\phi_1}{d\zeta} \frac{\delta(u_\perp)}{2\pi u_\perp} \frac{\delta(\sqrt{2\mathcal{E}} - u_B)}{u_B(\sqrt{2\mathcal{E}} - u_B)},$$

we obtain from Eq. (16)

(18)
$$\bar{f}_1^- = -\frac{\delta(u_\perp)}{2\pi u_\perp} \delta(\sqrt{2\mathcal{E}} - u_B)\zeta - \frac{\delta(u_\perp)}{2\pi u_\perp} \frac{\delta(\sqrt{2\mathcal{E}} - u_B)}{u_B(\sqrt{2\mathcal{E}} - u_B)} \phi_1(\zeta)$$

if $u_z > 0$ and $\mathcal{E} \geq 0$, and $\bar{f}_1^- = 0$ otherwise.

Since $\phi_0(\zeta)$ is a monotonically decreasing function, we transform the variable ζ to y by

(19)
$$y = -2\phi_0(\zeta).$$

Evidently $y \geq 0$ and $y \geq -2\mathcal{E}$. Then Eq. (15) becomes

(20)
$$\frac{\partial \bar{f}_1^+}{\partial y} = -\frac{\delta(\sqrt{h} - u_B)}{2\phi_0' \pi (2\mathcal{E} + y)}$$

if $u_z > 0$ and $h \geq 0$. Otherwise $\partial \bar{f}_1^+/\partial y = 0$. Here $\phi_0' = d\phi_0/d\zeta$ is evaluated at $\zeta = \zeta(y) = \phi_0^{-1}(-y/2)$. The function h may be written in terms of $\mathcal{E}$, $u_\perp$, and y as

$$h = 2(u_\perp^2 + \mathcal{E}) + \frac{u_\perp^4}{2\mathcal{E} + y},$$

which is a monotonically decreasing function of y ($> -2\mathcal{E}$). In integrating Eq. (15), we find

(21) $$\bar{f}_1^+ = \frac{u_B}{-\pi \phi_0'(\zeta_c)(u_B^2 - 2(u_\perp^2 + \mathcal{E}))}$$

if

(22) $$h = 2(u_\perp^2 + \mathcal{E}) + \frac{u_\perp^4}{2\mathcal{E} + y} \leq u_B^2$$

(23) and $$2(u_\perp^2 + \mathcal{E}) + \frac{u_\perp^4}{2\mathcal{E}} \geq u_B^2$$

on the domain $y \geq 0$, $y \geq -2\mathcal{E}$, and $u_\perp \geq 0$. Otherwise, $\bar{f}_1^+ = 0$. In Eq. (21), $\zeta_c = \zeta_c(u_\perp, \mathcal{E})$ denotes the ζ value that satisfies $h = u_B^2$, or $\zeta_c = \zeta(y_c)$ with

(24) $$y_c = -2\mathcal{E} + \frac{u_\perp^4}{u_B^2 - 2(u_\perp^2 + \mathcal{E})}.$$

3. Angular distributions.

We now calculate the angular distribution of the ion flux

$$\Gamma_\theta(\theta, z) = \int_0^{2\pi} d\varphi \int_0^\infty v^2 dv v_z f(v, \theta, z) \sin\theta$$

(25) $$= \frac{2\pi n_I \omega_{pi} d^2 \sin\theta \cos\theta}{\lambda_{mfp}} \int_0^\infty du u^3 \bar{f}_1^+(\mathcal{E}, u_\perp, \zeta),$$

where

(26) $$\mathcal{E} = \frac{1}{2}u^2 \cos^2\theta + \phi_0(\zeta) \quad \text{and} \quad u_\perp = u\sin\theta.$$

In this section, we are concerned only with angular distributions for $\theta > 0$ and do not count ballistic ions ($\theta = 0$) whose distribution function is given by the δ function.

In order to carry out the integration of Eq. (25), we need to determine the range of the integration variable u for which $\bar{f}_1^+$ is given by Eq. (21). From the inequality (22) and Eq. (26), we obtain the condition

(27) $$u^2 \leq (u_B^2 + y)\cos^2\theta.$$

Substituting Eq. (26) into the inequality (23) yields the condition

(28) $$u^4 - (2y + u_B^2 \cos^2\theta)u^2 + (y + u_B^2)y \geq 0.$$

The discriminant of this quadratic equation for u^2 is given by

$$D = (2y + u_B^2 \cos^2\theta)^2 - 4(y + u_B^2)y$$
$$= u_B^2(u_B^2 \cos^4\theta - 4y \sin^2\theta).$$

Therefore, the range of u is given as follows: if

(29) $$\frac{u_B^2}{4y} < \frac{\sin^2\theta}{\cos^4\theta} \quad (\Longleftrightarrow D < 0)$$

then we only need Eq. (27), i.e.,

(30) $$0 \leq u^2 \leq (u_B^2 + y)\cos^2\theta.$$

On the other hand, if

(31) $$\frac{u_B^2}{4y} \geq \frac{\sin^2\theta}{\cos^4\theta} \quad (\Longleftrightarrow D \geq 0)$$

then inequalities (27) and (28) must hold simultaneously, i.e.,

(32) $$0 \leq u^2 \leq y + \frac{u_B^2}{2}\cos^2\theta - \frac{1}{2}\sqrt{D}$$
$$\text{and} \quad y + \frac{u_B^2}{2}\cos^2\theta + \frac{1}{2}\sqrt{D} \leq u^2 \leq (u_B^2 + y)\cos^2\theta.$$

It is easy to show that $(u_B^2 + y)\cos^2\theta \geq y + (u_B^2/2)\cos^2\theta + \sqrt{D}/2$ for all $y \geq 0$ and θ. It should be noted that the term $u_B^2/y = mv_B^2/2q|\Phi_0|$ denotes the ratio of the initial ion kinetic energy to the zeroth–order potential energy Φ_0 at $z = \zeta d$ and typically takes a small value.

In the case $u_B^2/4y = mv_B^2/8q|\Phi_0| < \sin^2\theta/\cos^4\theta$ where the inequality (30) holds, therefore, the angular distribution of the ion flux is given by

(33) $$\Gamma_\theta(\theta, z) = \frac{2\pi n_I \omega_{pi} d^2 \sin\theta \cos\theta}{\lambda_{mfp}} \times$$
$$\int_0^{\sqrt{u_B^2+y}\cos\theta} \frac{u_B u^3 du}{-\pi\phi_0'(\zeta_c)((u_B^2 + y) - u^2(1 + \sin^2\theta))},$$

where the relation $u_B^2 - 2(u_\perp^2 + \mathcal{E}) = (u_B^2 + y) - u^2(1 + \sin^2\theta)$ is used. For small angles θ satisfying $u_B^2/4y \geq \sin^2\theta/\cos^4\theta$, the range of integration of Eq. (33) needs to be changed according to the inequality (32).

If the electric field is constant, then the term $-\phi_0' = \bar{E}_I$ may be taken outside of the integration and we can in fact carry out the integration:

(34) $$\Gamma_\theta(\theta, z) = \frac{d}{\lambda_{mfp}}\Gamma_0 \frac{(u_B^2 - 2\phi_0(\zeta))}{\bar{E}_I} G_\theta(\theta),$$

where $\Gamma_0 = n_I v_B$ denotes the total flux and

(35) $$G_\theta(\theta) = \frac{2\sin\theta\cos\theta}{(1+\sin^2\theta)^2}\left(4\log\frac{1}{\sin\theta} + \sin^4\theta - 1\right).$$

Equation (35) gives the profile of the ion flux angular distribution.

If the initial velocity v_B is sufficiently small, so that the condition $u_B^2/4y \ll 1$ is satisfied, then the inequality $u_B^2/4y = mv_B^2/8q|\Phi_0| < \sin^2\theta/\cos^4\theta$ holds for most of

$\theta > 0$, i.e., $u_B/2y \lesssim \theta \leq \pi/2$. In this case, Eq. (34) may be further simplified with the use of $u_B^2 \ll -\phi_0 = \bar{E}_I \zeta$ and given in dimensional form by

$$\Gamma_\theta(\theta, z) = \frac{\Gamma_0 z}{\lambda_{mfp}} \mathcal{G}_\theta(\theta). \tag{36}$$

Thus, Γ_θ is seen to be independent of the electric field strength E_I in the case of a constant electric field.

In the case of self–consistent electric fields, where the potential ϕ_0 is obtained by solving Eq. (10), the electric field ϕ'_0 is no longer independent of ζ. The angular distribution $\Gamma_\theta(\theta, z)$ then needs to be calculated directly from Eq. (33) with knowledge of the dependence of ϕ'_0 on u and θ.

Multiplying Eq. (10) by $d\phi_0/d\zeta$ and integrating with respect to ζ, we obtain

$$\frac{1}{2}\left(\frac{d\phi_0}{d\zeta}\right)^2 = -u_B \int_0^{\phi_0} \frac{d\phi_0}{\sqrt{u_B^2 - 2\phi_0}} + \frac{1}{2}\bar{E}_I^2,$$

where the boundary condition $d\phi_0/d\zeta = \bar{E}_I$ is used. Carrying out this integration and substituting $\phi_0 = -y_c/2$ yields

$$\phi'_0(\zeta_c) = -\sqrt{2(u_B\sqrt{u_B^2 + y_c} + u_B^2 K)}, \tag{37}$$

where $K = \bar{E}_I^2/2u_B^2 - 1$ (> -1). It is shown in [10] that the potential ϕ_0 is a weak function of K for realistic values of K ($-1 < K < \frac{1}{3}$). The function y_c given in Eq. (24) satisfies

$$u_B^2 + y_c = \frac{((u_B^2 + y) - u^2)^2}{(u_B^2 + y) - (1 + \sin^2\theta)u^2}. \tag{38}$$

Equations (37) and (38) give an expression for the dependence of $\phi'_0(\zeta_c)$ on u and θ.

Introducing

$$\xi = \frac{u}{u_{max}}, \quad \alpha = \frac{u_B}{u_{max}} \quad \text{with} \quad u_{max} = \sqrt{u_B^2 + y}, \tag{39}$$

we may write $-\phi'_0(\zeta_c) = \sqrt{2}\alpha u_{max} g_a(\xi)$, where

$$g_a(\xi) = \sqrt{\sqrt{(1-\xi^2)^2/(1-(1+\sin^2\theta)\xi^2)} + \alpha K}. \tag{40}$$

From Eq. (33), the angular distribution of the ion flux is then given by

$$\Gamma_\theta(\theta, z) = \Gamma_0 \frac{d}{\lambda_{mfp}} \frac{u_B}{\sqrt{2\alpha^3}} \mathcal{G}_\theta^{sc}(\theta; \alpha, K), \tag{41}$$

where, in the case of a self–consistent field, we define

$$\mathcal{G}_\theta^{sc}(\theta; \alpha, K) = 2\sin\theta\cos\theta \int_{l(\alpha)} \frac{\xi^3 d\xi}{g_a(\xi)(1 - (1+\sin^2\theta)\xi^2)}. \tag{42}$$

Here the range of integration $I(\alpha)$ is given as follows: from the inequality (30),

$$\text{if} \qquad \alpha < \frac{2\sin\theta}{(1+\sin^2\theta)} \qquad \left(\iff \frac{u_B^2}{4y} < \frac{\sin^2\theta}{\cos^4\theta}\right),$$

(43) $\qquad$ then $\qquad I(\alpha) = \{0 \leq \xi \leq \cos\theta\}.$

Although $u_B^2/4y$ is generally small and the angular distribution for most values of θ is given by the integration over $I(\alpha)$ above, an accurate account of the small-angle distribution must be given by a different integration range $I(\alpha)$. From the inequality (32),

$$\text{if} \qquad \alpha \geq \frac{2\sin\theta}{(1+\sin^2\theta)} \qquad \left(\iff \frac{u_B^2}{4y} \geq \frac{\sin^2\theta}{\cos^4\theta}\right),$$

(44) $\qquad$ then $\qquad I(\alpha) = \begin{cases} 0 \leq \xi \leq \left((1-\alpha^2) + \frac{\alpha^2}{2}\cos^2\theta - \frac{\hat{D}^{\frac{1}{2}}}{2}\right)^{\frac{1}{2}}, \\ \left((1-\alpha^2) + \frac{\alpha^2}{2}\cos^2\theta + \frac{\hat{D}^{\frac{1}{2}}}{2}\right)^{\frac{1}{2}} \leq \xi \leq \cos\theta. \end{cases}$

Here $\hat{D} = \alpha^2(\alpha^2\cos^4\theta - 4(1-\alpha^2)\sin^2\theta)$. We note that the function $\mathcal{G}_\theta^{sc}$ depends on K through the term αK in the function $g_a(\xi)$ of Eq. (40). Since the value α is typically small and the dependence of the potential ϕ_0 on the parameter K is known to be weak [10], the parameter dependence of $\mathcal{G}_\theta^{sc}$ on K is also weak.

Figure 1 shows a comparison of the theoretically-predicted angular distributions and Monte Carlo simulation results in the case $d/\lambda_{mfp} = 0.14$. The electric field used in the Monte Carlo simulation and the self-consistent-field distribution (Eq. (42), the solid line) is the solution to Eq. (10), subject to the boundary conditions $\phi_0(0) = 0$ and $\bar{E}_I = d\phi_0(0)/d\zeta = 2.8 \times 10^{-4}$. The constant-field approximation (Eq. (35)) is given by the dashed line. The theoretical distribution for the ballistic ion component, which is a delta function at $\theta = 0$, is not shown here and the Monte Carlo ballistic ion component, represented by the first bin at $\theta = 0$, is truncated by the frame of the figure; all curves are normalized so as to enclose unit area. A good agreement between the analytic distributions and the simulation results is evident in Fig. 1.

4. Energy distributions. The energy distribution of the ion flux $\Gamma_{EN}(\eta, z)$ is defined as

(45) $\qquad \Gamma_{EN}(\eta, z)d\eta = \int_0^{2\pi} d\varphi \int_0^\pi \sin\theta d\theta v_z f v^2 dv,$

where η denotes the ratio of the ion kinetic energy to the kinetic energy of the ballistic ions at z, i.e.,

(46) $\qquad \eta = \frac{\frac{1}{2}mv^2}{\frac{1}{2}mv_B^2 - q\Phi_0} = \frac{u^2}{u_B^2 + y}.$

For scattered ions ($\eta < 1$), we have from Eq. (45)

(47) $\qquad \Gamma_{EN}(\eta, z) = \frac{n_I(u_B^2 + y)d}{\lambda_{mfp}} \int_{J(\alpha)} \frac{d\theta}{-\phi_0'(\zeta_c)(1 - (1+\sin^2\theta)\eta)}.$

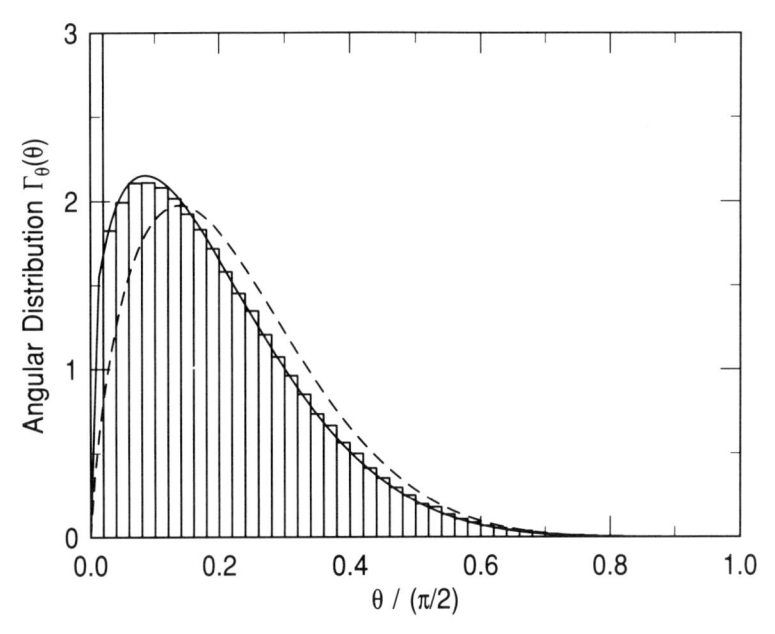

FIG. 1. *The angular distributions of the ion flux in the case of a self-consistent electric field obtained from the Monte Carlo simulations (histogram) and Eq. (42) (the solid curve). For comparison, the formula for the constant-field approximation (Eq. (35)) is also presented as a dashed line. The dimensionless parameters used here are* $d/\lambda_{mfp} = 0.14$, $u_B = 1.0 \times 10^{-2}$, $\bar{E}_I = 5.1 \times 10^{-3}$, *and* $\zeta = 1$.

Here we have used the relation $u_B^2 - 2(u_1^2 + \mathcal{E}) = (u_B^2 + y)(1 - (1 + \sin^2\theta)\eta)$.

The range of integration $J(\alpha)$ for θ is obtained from the conditions (27) and (28): If $0 \leq \eta \leq 1 - \alpha^2$, then

(48) $$\eta \leq \cos^2\theta \leq 1,$$

and if $1 - \alpha^2 \leq \eta < 1$, then

(49) $$\eta \leq \cos^2\theta \leq \frac{\eta^2 + (1 - 2\eta)(1 - \alpha^2)}{\eta\alpha^2}.$$

In the case of a constant electric field, we may substitute $-\phi_0'(\zeta_c) = \bar{E}_I$ into Eq. (47). We thus obtain

(50) $$\Gamma_{EN}(\eta, z) = \Gamma_0 \frac{z}{\lambda_{mfp}} \frac{1}{(1 - \alpha^2)} \mathcal{G}_{EN}(\eta; \alpha),$$

where

(51) $$\mathcal{G}_{EN}(\eta; \alpha) = \begin{cases} -\log(1 - \eta) & (\eta \leq 1 - \alpha^2) \\ -\log \alpha^2 & (1 - \alpha^2 < \eta < 1), \end{cases}$$

which holds for any α ($0 < \alpha < 1$). We note that the distribution is constant for $1 - \alpha^2 < \eta < 1$.

In the case of a self-consistent electric field, the potential ϕ_0 is obtained by solving Eq. (10). In this case, the electric field ϕ_0' also becomes a function of η and θ. As shown in Eqs. (37) and (38), we may write $-\phi_0'(\zeta_c) = \sqrt{2\alpha} u_{max} g_E(\theta)$, where

(52) $$g_E(\theta) = \sqrt{\sqrt{(1 - \eta)^2/(1 - (1 + \sin^2\theta)\eta)} + \alpha K}.$$

Here the function $g_E(\theta)$ and the function $g_a(\xi)$ of Eq. (40) are equivalent, with the relation $\eta = \xi^2 = u^2/(u_B^2 + y)$. From Eq. (47), the energy distribution of the ion flux for a self-consistent electric field is then given by

(53) $$\Gamma_{EN}(\eta, z) = \Gamma_0 \frac{d}{\lambda_{mfp}} \frac{u_B}{\sqrt{2\alpha^3}} \mathcal{G}_{EN}^{sc}(\eta; \alpha, K),$$

where

(54) $$\mathcal{G}_{EN}^{sc}(\eta; \alpha, K) = \eta \int_{J(\alpha)} \frac{\sin\theta \cos\theta \, d\theta}{g_E(\theta)(1 - (1 + \sin^2\theta)\eta)}.$$

Figure 2 shows a comparison of theoretically-predicted energy distributions (the constant-field approximation (51) for the dashed line and the self-consistent-field distribution (54) for the solid line) to Monte Carlo simulation results in the case $d/\lambda_{mfp} = 0.14$. The electric field used in the Monte Carlo simulations and for Eq. (54) is the same as that used in Fig. 1. The analytical distribution for the ballistic ion component, which is a delta function at $\eta = 1$, is not shown in Fig. 2 and the Monte Carlo ballistic ion component, represented by the bin at $\eta = 1$ is truncated by the frame of the figure. A good agreement between the analytical formulae and the Monte Carlo results is clearly seen in Fig. 2.

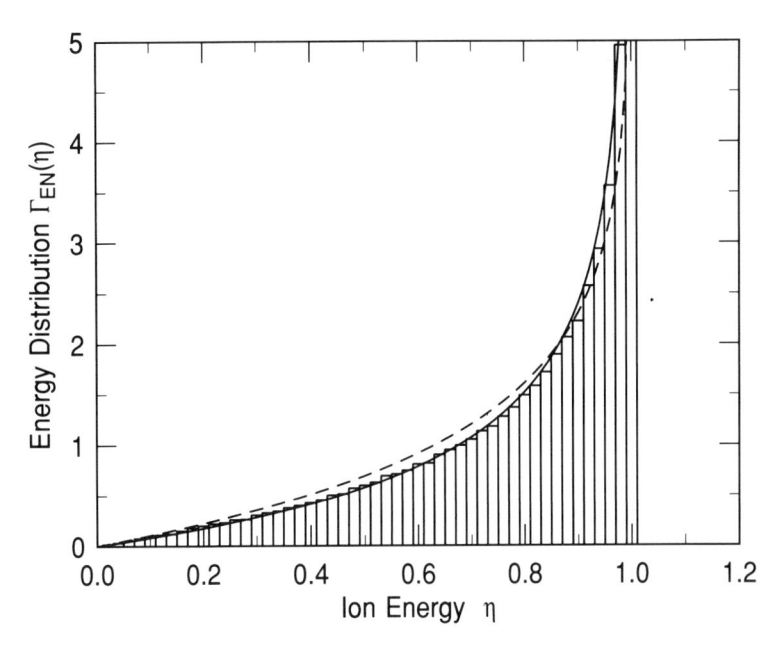

FIG. 2. *The energy distributions of the ion flux in the case of a self-consistent electric field obtained from the Monte Carlo simulations (histogram) and Eq. (54)(the solid curve). For comparison, the formula for the constant-field approximation (Eq. (51)) is also presented as a dashed line. The dimensionless parameters used here are the same as those for Fig. 1.*

5. RF sheaths. We now briefly discuss ion distributions in RF sheaths. For the sake of simplicity we consider only a collisionless sheath, to which a time–periodic cathode voltage is applied. We also assume, as before, the electrons to be sufficiently depleted in the sheath that the sheath electric field is determined only by the ion space charge. The validity of these assumptions will be discussed later in this section. The ion velocity distribution function $f(t, z, v_z)$ at position z is then governed, after some normalization, by the following Vlasov–Poisson system:

$$(55) \qquad \frac{\partial \bar{f}}{\partial \tau} + u_z \frac{\partial \bar{f}}{\partial \zeta} - \varepsilon^2 \frac{d\phi}{d\zeta} \frac{\partial \bar{f}}{\partial u_z} = 0,$$

$$(56) \qquad \frac{d^2\phi}{d\zeta^2} = - \int \bar{f}(\tau, \zeta, u_z) du_z,$$

where the potential satisfies a time–periodicity, i.e., $\phi(\tau, \zeta) = \phi(\tau + 2\pi, \zeta)$. The boundary condition for $\bar{f}(\tau, \zeta, u_z)$ is given by

$$(57) \qquad \bar{f}(\tau, \zeta = 0, u_z) = \delta(u_z - u_B),$$

namely, the ions are assumed to be injected into the sheath as a beam in the z direction with velocity u_B. Since the sheath is collisionless, the perpendicular component of the velocity vanishes, i.e., $u_\perp = 0$, and we may consider that the distiribution function $\bar{f}$ is already integrated over $u_\perp$. The normalizations used here are thus somewhat different from those in Eq. (3) and are given by

(58)
$$\tau = \omega t, \quad u_z = v_z/\omega d,$$
$$f = \bar{f} n_I/(\omega d), \quad \varepsilon = \omega_{pi}/\omega,$$

where $\zeta = z/d$ and $\Phi = \phi q n_I d^2/\varepsilon_0$ are the same as those in Eq. (3), and $\omega_{pi} = q\sqrt{n_I/m\varepsilon_0}$ denotes the ion plasma frequency. The boundary conditions for the potential $\phi(t,z)$ are given by

(59)
$$\phi(\tau, 0) = 0 \quad \text{and} \quad \phi(\tau, \zeta = 1) = \bar{V}(\tau),$$

where $\bar{V}(\tau) = \varepsilon_0 V(\omega t)/(q n_I d^2)$ denotes the normalized cathode voltage. Here $V(\omega t)$ is given as a time–periodic function in t with period $2\pi/\omega$.

The time average of a function $h(\tau)$ is defined as

(60)
$$\langle h \rangle = \frac{1}{2\pi} \int_0^{2\pi} h(\tau) d\tau.$$

We also introduce the following time–integral operator $\mathcal{L}$ applied to a time–dependent function $h(\tau)$:

$$\mathcal{L} h = \int_0^{\tau} h(\tau') d\tau' - \left\langle \int_0^{\tau} h(\tau') d\tau' \right\rangle.$$

If the function $h(\tau)$ is 2π–periodic in τ and satisfies $\langle h \rangle = 0$, then the function $\mathcal{L}h$ is also 2π–periodic in τ and satisfies $\langle \mathcal{L}f \rangle = 0$. Defining $\phi_0(\zeta) = \langle \phi \rangle$ and $\phi_1(\tau, \zeta) = \phi - \langle \phi \rangle$, we obtain the characteristic equations for Eq. (55):

(61)
$$\frac{d\zeta}{d\tau} = u_z,$$

(62)
$$\frac{du_z}{d\tau} = -\varepsilon^2 \frac{\partial}{\partial \zeta}[\phi_0(\zeta) + \phi_1(\tau, \zeta)].$$

We now seek the time–periodic solution of the system (55) and (56) satisfying

$$\bar{f}(\tau, \zeta, u_z) = \bar{f}(\tau + 2\pi, \zeta, u_z),$$

in the regime of high frequency, i.e., $\varepsilon \ll 1$. When $\varepsilon \ll 1$, the solution to the characteristic equations at z is given [11], up to order ε^2, by

(63)
$$u_z = \left[u_B^2 - 2\varepsilon^2 \phi_0(\zeta)\right]^{\frac{1}{2}} - \varepsilon^2 \mathcal{L}\phi_1' + \mathcal{O}(\varepsilon^3),$$

where $\phi_1'(\tau, \zeta) = \partial \phi_1/\partial \zeta$. The derivation of Eq. (63) is based on a two–time scale asymptotic expansion in small ε. Since $d\bar{f}/d\tau = 0$ or $\bar{f}$ = constant along the characteristics, the standard method of characteristics yields the velocity distribution function at ζ:

(64)
$$\bar{f}(\tau, \zeta, u_z) = \delta\left(\left\{[u_z + \varepsilon^2 \mathcal{L}\phi_1']^2 + 2\varepsilon^2 \phi_0(\zeta)\right\}^{\frac{1}{2}} - u_B\right)$$
$$= \frac{u_B}{u_{0f}(\zeta)} \delta\left(u_z - [u_{0f}(\zeta) - \varepsilon^2 \mathcal{L}\phi_1']\right),$$

where $u_{0f}(\zeta) = [u_B^2 - 2\varepsilon^2 \phi_0(\zeta)]^{\frac{1}{2}}$. Here, terms up to order ε^2 only in Eq. (63) are used to derive Eq. (64).

Writing the kinetic energy as $\mathcal{E} = mv_z^2/2$, the energy distribution $\Gamma_{\mathcal{E}}$ of the ion flux at the cathode is given by $\Gamma_{\mathcal{E}} d\mathcal{E} = v_z f(t, d, v_z) dv_z$, i.e., $\Gamma_{\mathcal{E}} = f(t, d, v_z)/m$. The experimentally-observed ion energy flux distribution is then the time average of $\Gamma_{\mathcal{E}}$:

$$\langle \Gamma_{\mathcal{E}} \rangle = \frac{\omega}{2\pi} \int_0^{2\pi/\omega} \Gamma_{\mathcal{E}} \, dt = \frac{n_I}{2\pi m \omega d} \int_0^{2\pi} \bar{f}(\tau, \zeta, u_z) \, d\tau. \tag{65}$$

In dimensional form, we obtain from Eq. (64)

$$\langle \Gamma_{\mathcal{E}} \rangle (\mathcal{E}) = \frac{n_i v_I}{\pi v_{0f}} \sqrt{\frac{m\mathcal{E}}{2}} \sum_i \left| \frac{d\mathcal{E}_\varphi}{d\varphi} \right|_{\varphi_i}^{-1}, \tag{66}$$

where $v_{0f} = u_{0f} \omega d$, $\mathcal{E}_\varphi$ denotes the final kinetic energy as a function of the final phase $\varphi = \omega t$, i.e.,

$$\mathcal{E}_\varphi = \frac{1}{2} m v_f^2 \simeq \frac{1}{2} m v_{0f}^2 + \frac{v_{0f} q}{\omega} \mathcal{L} E_1 \big|_{z=d}, \tag{67}$$

and the discrete phases $\varphi_i = \varphi_i(\mathcal{E})$ denote all the distinct solutions φ of the equation $\mathcal{E} = \mathcal{E}_\varphi(\varphi)$. In the case of sinusoidal time dependence i.e. $E(t, z) = -\partial \Phi / \partial z = E_0(z) + E_1(z) \cos \omega t$, the general expression (66) reduces to

$$\langle \Gamma_{\mathcal{E}} \rangle (\mathcal{E}) = \frac{n_i v_I}{\pi m v_{0f} \sqrt{\left(v_+ - \sqrt{2\mathcal{E}/m}\right)\left(\sqrt{2\mathcal{E}/m} - v_-\right)}}, \tag{68}$$

where

$$v_{\pm} = v_{0f} \pm \frac{q E_1(d)}{m\omega}.$$

Numerical calculations of the energy distribution $\langle \Gamma_{\mathcal{E}} \rangle$, based on Eq. (68) and Monte Carlo simulations, are found in [11].

Equation (64) gives the ion velocity distribution for any given electric field potential ϕ. However, in order to obtain a self-consistent electric field profile, one must solve the Poisson equation (56), using Eq. (64). Carrying out the integration of Eq. (56), we obtain

$$\frac{d^2 \phi}{d\zeta^2} = -\frac{u_B}{\sqrt{u_B^2 - 2\varepsilon^2 \phi_0(\zeta)}}. \tag{69}$$

It is easy to see that Eq. (69) may be split into the following two equations:

$$\frac{d^2 \phi_0}{d\zeta^2} = -\frac{u_B}{\sqrt{u_B^2 - 2\varepsilon^2 \phi_0(\zeta)}}, \tag{70}$$

$$\frac{d^2 \phi_1}{d\zeta^2} = 0. \tag{71}$$

Suppose that the (normalized) time-dependent cathode voltage is given by $\bar{V}(\tau) = \bar{V}_0 + \bar{V}_1 \cos \tau$. Then the boundary conditions for Eqs. (70) and (71) become

$$\phi_0(0) = 0, \quad \phi_0(1) = \bar{V}_0, \tag{72}$$

and

(73) $$\phi_1(0) = 0, \quad \phi_1(1) = \bar{V}_1 \cos\tau.$$

Equation (70) subject to the boundary conditions (72) has a form similar to Eq. (10), giving a collisionless DC–sheath potential. Equation (71) with the boundary conditions (73) gives a uniform (i.e., z–independent) oscillation field, i.e., $\partial\phi_1/\partial\zeta = \bar{V}_1 \cos\tau$.

The z–independent oscillating electric field derived above, however, is physically implausible since in real RF sheaths the high–mobility electron gas responds almost instantaneously to the RF excitation and gives rise to a rapid oscillating motion of the relatively sharp presheath/sheath boundary. A realistic self-consistent model must incorporate such electron effects in Poisson's equation (56). For example, if the electrons are governed by the Boltzmann distribution (i.e., the electron density is proportional to $\exp(q\phi/k_B T_e)$), then Poisson's equation shoud be given, instead of Eq. (56), by

(74) $$\frac{d^2\phi}{d\zeta^2} = -\int \bar{f}(\tau,\zeta,u_z)du_z + \exp\left(\frac{d^2}{\lambda_d^2}\phi\right),$$

where $\lambda_d = (\varepsilon_0 k_B T_e / n_I q^2)^{\frac{1}{2}}$ denotes the Debye length. The self–consistent solution from Eqs. (64) and (74) is beyond our present scope and will be discussed elsewhere. Discussion of the effects of a high–frequency motion of the sharp presheath/sheath boundary on the ion distribution may be found in [12].

6. Discussion. From the steady state Boltzmann–Poisson system, we derive analytic formulae for the angular and energy distributions of the incident ion flux at the cathode surface in weakly–collisional DC sheaths. The resulting distributions, given by Eqs. (35), (42), (51) and (54), are compared with Monte Carlo simulations and are found to be in good agreement. Although the model considered here is an idealization of a real DC sheath, it captures the essential physics of the collision mechanism and the self-consistent electric field. The idealization, moreover, renders the model amenable to a relatively simple analytical treatment, and clarifies the dependence of the distributions on various plasma parameters. In fact, it is not difficult to incorporate detailed physical features, such as more realistic presheath/sheath boundary conditions and energy-dependent collision-cross sections, in the Monte Carlo simulations in order to compare the resulting ion distributions with experimental results under various conditions.

As for RF discharges, we briefly discussed the ion distribution in collisionless RF sheaths, based on the time–periodic Vlasov–Poisson system. The time–averaged energy distribution of incident ion flux at the cathode was derived analytically in Eq. (68) for a given sinusoidally oscillating electric field. Using a more general expression for the distribution function given in Eq. (64), we obtain a simple configuration of self–consistent RF electric field without taking into account the high–mobility electron effects. However, more realistic self-consistent RF fields (taking into account the highly mobile electron gas) could differ significantly from the RF field discussed in the present work. A detailed discussion of this matter is beyond our present scope and we defer it to a future publication.

REFERENCES

[1] S. M. Sze, *VSLI Technology*, McGraw–Hill, New York (1988).
[2] B. Chapman, *Glow Discharge Processes*, John Wiley & Sons, New York (1980).
[3] J. W. Coburn and E. Kay, J. Appl. Phys. **43**, 4965 (1972).
[4] W. M. Holber and J. Forster, J. Vac. Sci. Technol. **A8**, 3720 (1990).
[5] R. T. Farouki, S. Hamaguchi, and M. Dalvie, Phys. Rev. A **44**, 2664 (1991).
[6] S. Hamaguchi, R. T. Farouki, and M. Dalvie, Phys. Rev. A **44**, 3804 (1991).
[7] See, for example, L. D. Landau and E. M. Lifshitz, *Mechanics*, Pergamon Press, Oxford, 1960.
[8] W. P. Allis, "Motion of Ions and Electrons," in *Handbuch der Physik*, Vol. 21, Springer-Verlag, Berlin, 1956, p. 383.
[9] C. D. Child, Phys. Rev. **32**, 492 (1911): I. Langmuir, Phys. Rev. (Ser. II) **2**, 450 (1913).
[10] R. T. Farouki, M. Dalvie, and L. F. Pavarino, J. Appl. Phys. **68**, 6106 (1990).
[11] S. Hamaguchi, R. T. Farouki, and M. Dalvie, Phys. Rev. Lett. **68**, 44 (1992).
[12] R. T. Farouki, S. Hamaguchi, and M. Dalvie, Phys. Rev. A**45**, 5913 (1992)

AN INTERFACE METHOD FOR SEMICONDUCTOR PROCESS SIMULATION

MICHAEL J. JOHNSON* AND CARL L. GARDNER**

Abstract. The diffusion of dopants in silicon at high temperatures is modeled by a nonlinear parabolic system of partial differential equations on a two-dimensional region with a moving boundary. A numerical solution using the L-stable TRBDF2 time integration method and a "box method" spatial discretization is described.

Details are given of the methods used to specify and manipulate curves, and to define arbitrary simply connected regions by their boundary curves. Numerical experiments are presented comparing the divided difference and TR/TR methods for dynamically adjusting the timestep, and comparing Newton and Newton-Richardson iteration.

1. Introduction.

Semiconductor process simulation[1] models the nonlinear diffusion of dopant atoms during the thermal annealing of silicon or other semiconductor wafers which have been doped by ion implantation. When the temperature is raised, the dopant atoms diffuse in the sample. The diffusivities of the dopant atoms depend on local dopant concentration and change with time due to a number of transient effects. During the anneal, portions of the surface of the wafer are allowed to oxidize. The rate of oxide growth depends in part on local dopant concentration along the oxide/silicon boundary, so that the mathematical model of this process becomes a free boundary problem.

The nonlinear diffusion process is described by a set of conservation laws for impurities

$$(1) \qquad \frac{\partial C_\alpha}{\partial t} = \nabla \cdot \left(\sum_\beta D_{\alpha\beta} \nabla C_\beta \right) \equiv F_\alpha$$

in a simply connected region $\Omega(t)$, where $C_\alpha(\mathbf{x}, t)$ is the concentration of the αth species of dopant, D is a matrix of phenomenological diffusion coefficients, and α, $\beta = 1, \ldots, N$ label the types of impurities. Note that $D = D(C_1, \ldots, C_N)$ includes the effects of the coupling of the impurity ions to the electric field.

The boundary $\partial\Omega(t)$ of $\Omega(t)$ represents the union of a silicon/mask interface and a silicon/oxide interface. Boundary conditions of homogeneous Neumann type are imposed by the physical constraints that no dopant ions may leave the silicon region $\Omega(t)$ unless consumed by the growing oxide regions and that no migration occurs across the oxide/silicon interface:

$$(2) \qquad (\hat{n} \cdot \nabla C_\alpha)_{\partial\Omega(t)} = 0.$$

The code which positions the moving boundary is distinct from the code that calculates dopant diffusion. This paper addresses the efficient numerical solution of the diffusion problem only, using a prescribed boundary $\Omega(t)$.

---

*IBM Corporation, Endicott, NY, 13760.
**Department of Computer Science, Duke University, Durham, NC 27706. Research supported in part by the National Science Foundation under grant DMS-8905872.
[1] See Ref. [1] for a review.

2. Numerical methods.

We use the composite TRBDF2 method [2] to integrate the solution in time. To integrate Eq. (1) from $t = t_n$ to $t_{n+1} = t_n + \Delta t_n$, we first apply the trapezoidal rule (TR) to advance the solution from t_n to $t_{n+\gamma} = t_n + \gamma \Delta t_n$:

$$(3) \qquad C^{n+\gamma} - \gamma \frac{\Delta t_n}{2} F^{n+\gamma} = C^n + \gamma \frac{\Delta t_n}{2} F^n ,$$

and then use the second-order backward differentiation formula (BDF2) to advance the solution from $t_{n+\gamma}$ to t_{n+1}:

$$(4) \qquad C^{n+1} - \frac{1-\gamma}{2-\gamma} \Delta t_n F^{n+1} = \frac{1}{\gamma(2-\gamma)} C^{n+\gamma} - \frac{(1-\gamma)^2}{\gamma(2-\gamma)} C^n .$$

This composite one-step method is second-order accurate and L-stable [2]. The importance of L-stability for diffusion is illustrated for a 1D computation in Figure 1. After a single timestep with $\Delta t = 50 \Delta t_{\text{Euler}} = 50 \Delta y^2 / 2 D_{\max}$, the TR method, which is A-stable but not L-stable, exhibits severe unphysical oscillations near the maximum of C.

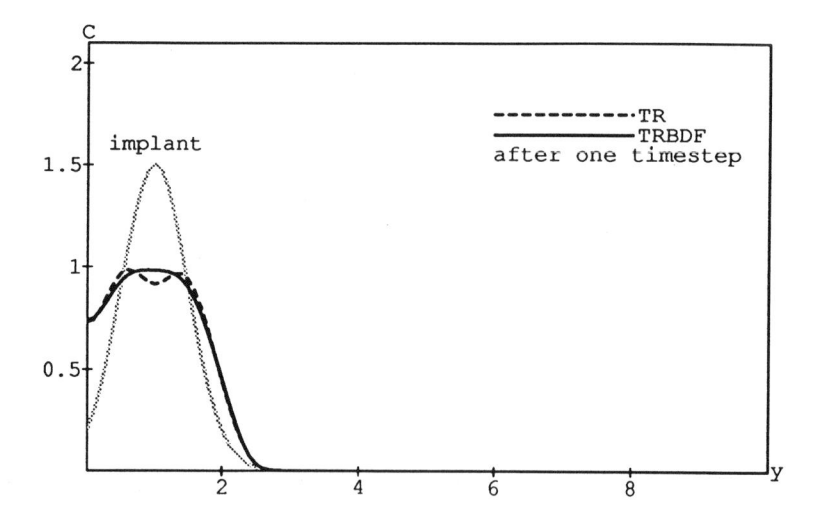

Figure 1: L-stable vs. A-stable (1D example).

We linearize F^{n+1} in Eq. (4) (and similarly $F^{n+\gamma}$ in Eq. (3)) by approximating

$$(5) \qquad F^{n+1}_{(k+1)} = F^{n+1}_{(k)} + \left(\frac{\delta F}{\delta C} \right)^{n+1}_{(k)} \delta C^{n+1}_{(k)}$$

where $k = 0, 1, \ldots$ labels the Newton iterations, and the Fréchet derivative

$$(6) \qquad \frac{\delta F}{\delta C} * = \nabla \cdot \left(\frac{\delta D}{\delta C} * \nabla C \right) + \nabla \cdot (D \nabla *) .$$

The new solution is obtained by setting

(7) $$C^{n+1}_{(k+1)} = C^{n+1}_{(k)} + \lambda \delta C^{n+1}_{(k)}, \quad C^{n+1}_{(0)} = C^{n+\gamma}$$

where λ is a damping factor [3] between 0 and 1, chosen to insure that the norm of the residual for Eq.(3) or (4) decreases monotonically. At each TR or BDF2 partial step, we iterate until the Newton method converges.

The Newton equation for the TR partial step is

(8)
$$\left[1 - \gamma\frac{\Delta t_n}{2}\left(\frac{\delta F}{\delta C}\right)^{n+\gamma}_{(k)}\right]\delta C^{n+\gamma}_{(k)} = -(C^{n+\gamma}_{(k)} - C^n) + \gamma\frac{\Delta t_n}{2}(F^{n+\gamma}_{(k)} + F^n) \equiv -G_{\text{TR}}$$

where G_{TR} is the residual for Eq. (3).

The Newton equation for the BDF2 partial step is

(9)
$$\left[1 - \frac{1-\gamma}{2-\gamma}\Delta t_n\left(\frac{\delta F}{\delta C}\right)^{n+1}_{(k)}\right]\delta C^{n+1}_{(k)} =$$
$$-\left(C^{n+1}_{(k)} - \frac{1}{\gamma(2-\gamma)}C^{n+\gamma} + \frac{(1-\gamma)^2}{\gamma(2-\gamma)}C^n\right) + \frac{1-\gamma}{2-\gamma}\Delta t_n F^{n+1}_{(k)} \equiv -G_{\text{BDF2}}$$

where G_{BDF2} is the residual for Eq. (4).

Note that the Jacobians for the TR and BDF2 partial steps have the same form if $\gamma = 2 - \sqrt{2}$. If the solution is varying slowly, then Jacobian factorizations may be reused (Newton-Richardson method) while retaining quadratic convergence of the Newton method [2]. The effectiveness of reusing Jacobians in semiconductor process simulations is discussed in Section 4.

Equations (8) and (9) are discretized in space by using the box method [4,5]. The box method, based on Gauss's theorem, evaluates the average divergence of a vector quantity **f** over a box as

(10) $$(\nabla \cdot \mathbf{f})_{\text{average, interior of box}} = \frac{\int_{\text{boundary of box}} \mathbf{f} \cdot \mathbf{n}\, ds}{\text{area of box}}.$$

The box method may be most easily implemented in computations on a rectangular grid by taking the sides of the boxes to be halfway between grid points. For example, along side 1 in Figure 2,

(11) $$(\nabla u) \cdot \mathbf{n} \approx \frac{u(b) - u(a)}{x(b) - x(a)},$$

and

(12) $$(v\nabla u) \cdot \mathbf{n} \approx \left(\frac{v(a) + v(b)}{2}\right)\left(\frac{u(b) - u(a)}{x(b) - x(a)}\right).$$

All spatial operators employed in the Newton method described above are of this type.

To define the box at the boundary $\partial\Omega(t)$, consider, for a moment, point a in Figure 2 as the origin of a coordinate system, and the dashed box as a unit square. In our implementation of the box method, the smallest incremental area is one octant of this unit square. To enable correct identification of interior octants, the list of boundary points adheres to an orientation convention such that the *previous* boundary point defines a starting octant, and the *next* boundary point defines a stopping octant, with the interior octants identified by counterclockwise rotation from *previous* to *next*, as shown in Figure 3.

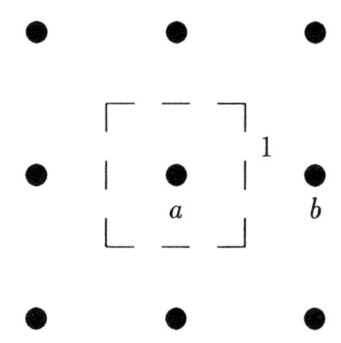

Figure 2: Box method in the interior of a rectangular grid.

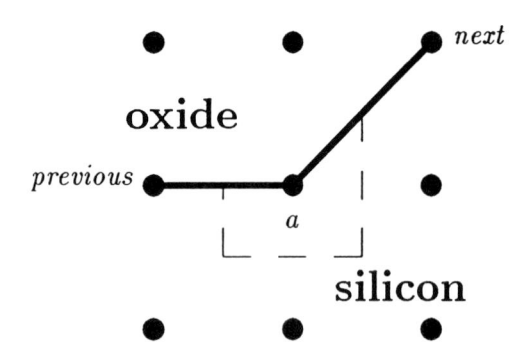

Figure 3: Box method on the boundary of a region.

The box method with central differences couples the solution at one point to the solution at nearby neighbors; there is no coupling between distant points. As a result, the matrix representing the spatially discretized operator (and consequently, the matrix to be solved at each timestep) is sparse. The discretized linear systems are solved using the sparse matrix package of Bank [2].

The timestep size Δt is adjusted dynamically within a window $[\Delta t_{\min}, \Delta t_{\max}]$ by monitoring a divided-difference estimate of the local truncation error τ [2]:

(13) $$\tau^{n+1} = k\Delta t_n^3 C^{(3)}$$

(14) $$\approx 2k\Delta t_n \left(\frac{1}{\gamma}F^n - \frac{1}{\gamma(1-\gamma)}F^{n+\gamma} + \frac{1}{1-\gamma}F^{n+1}\right),$$

where

(15) $$k = \frac{-3\gamma^2 + 4\gamma - 2}{12(2-\gamma)}.$$

The three values of F employed in Eq. (14) have already been calculated in the most recent TRBDF2 timestep.

An alternative approximation for τ^{n+1} involves re-taking the most recent partial timestep (from $t_{n+\gamma}$ to t_{n+1}) using TR instead of BDF2 [6]. (We will refer to the resulting value of C as $C_{TR/TR}^{n+1}$.) The TR/TR step yields the approximation

(16) $$\tau^{n+1} \approx 2.28(C_{TRBDF2}^{n+1} - C_{TR/TR}^{n+1})$$

for $\gamma = 2 - \sqrt{2}$. Very few Newton iterations are necessary in taking the second TR step, since C^{n+1} is, in fact, already known. The performances of the TR/TR and divided difference error estimators are compared in Section 4.

3. Interface method for moving boundaries.

The interface software which handles the geometry of regions with moving boundaries consists of a library of subroutines which define and manipulate certain data structures. The choice of structures employed here is motivated by the front-tracking method of Glimm and McBryan and coworkers (see, e.g., [7]). The method we describe is a simplified front-tracking code appropriate for *stable* interfaces. The major structures employed here are the **grid, curve, point, curve-point,** and **region** structures.

The grid structure defines the resolution in space. Included in the grid structure are lists of x and y coordinates, allowing a tensor product gridding; that is, variable spacing in either the x or y direction, or both. The goal of tensor product gridding is to increase overall accuracy by refining the mesh in certain areas [8].

The curve structure is a linked list of curve-point structures; each curve-point structure contains, in addition to its own x and y coordinates, a link to the previous and to the next curve-point structures. Curve-points need not lie on grid points. Figure 4 shows graphically the information contained in the curve and grid structures. The filled circles represent curve-point structures. It should be noted that the sinusoidal curve, used here as an example, is not indicative of a realistic oxide/silicon interface.

Each region structure defines one simply connected region by its boundary. Boundary curves are represented by curve structures which close on themselves but which are otherwise not self-intersecting. Within the region structure are tables assigning to each point in the region an index into the list of unknowns to be determined by the Newton method.

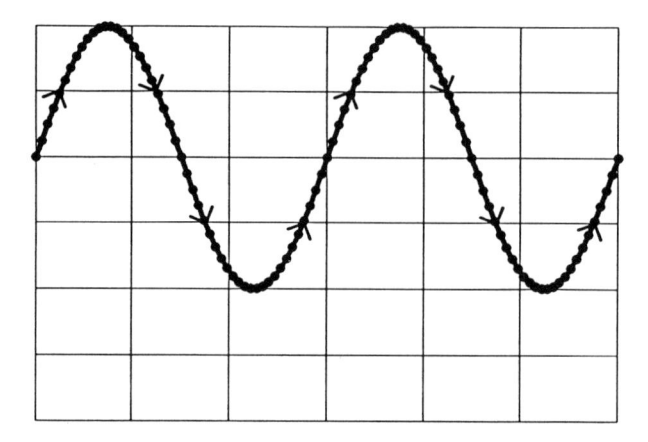

Figure 4: Graphical representation of curve and grid.

3.1. Processing of boundary curves. Figure 5 shows the curve of Figure 4 after additional points have been appended to construct a boundary curve.

The algorithm to define the interior of a region given its boundary is as follows. For each value of y on the grid, we note each x coordinate at which the boundary crosses this ($y = $ constant) line. Starting from the first such crossing, we mark every grid point as "in" until the second crossing, then every grid point is "out" until the third crossing, etc. By retaining (within the region structure) the minimum and maximum values of x and y as the region's boundary curve is formed, we avoid having to loop over every grid point when defining the interior of the transition region. Obviously, in order to follow this algorithm, the points of intersection of the the boundary curve with the grid lines must be known.

The process of finding these points of intersection is called **meshing** the curve. The algorithm for meshing a curve is complicated by the consideration of finite precision arithmetic; for example, a calculation which in exact arithmetic would produce, say, b slightly larger than a, might with truncation produce just the opposite. To avoid problems of this kind, at every step we adjust values which are very close to grid lines, moving them onto the line. More precisely, special variables called *smallx* and *smally* are established. Points within $2 \times$ *smallx* of a vertical grid line are initially moved to the line (and similarly for horizontal grid lines using *smally*). Thereafter, any two points within *smallx* of each other are considered to have the same x coordinate (likewise for y coordinates using *smally*).

After the adjustment, points are investigated in pairs. For each pair of points *(point1, point2)*, the mesh subroutine must determine whether a horizontal or vertical grid line has been crossed. If more than one grid line has been crossed, the crossing which is closest to *point1* is entered into the linked list of points in the meshed curve. This point of intersection then becomes the new *point1*. After all such grid crossings have been treated, *point2* is copied into the meshed curve if it is on a grid line (after adjustment). Then, *point2* becomes the new *point1*, and a new *point2* is extracted from the original curve. Figure 6 shows the curve of Figure 5 after meshing.

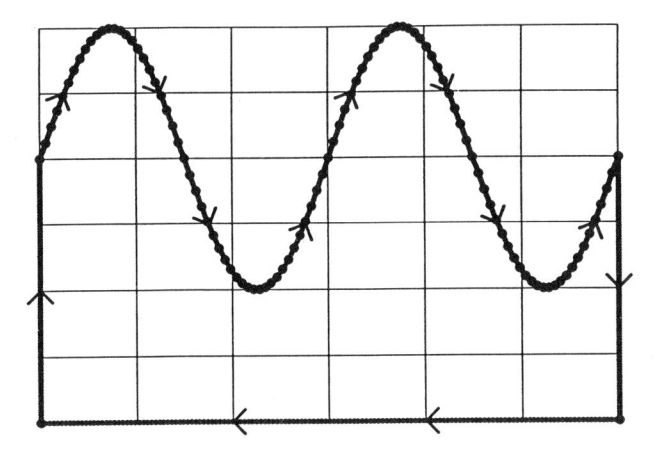

Figure 5: Boundary curve.

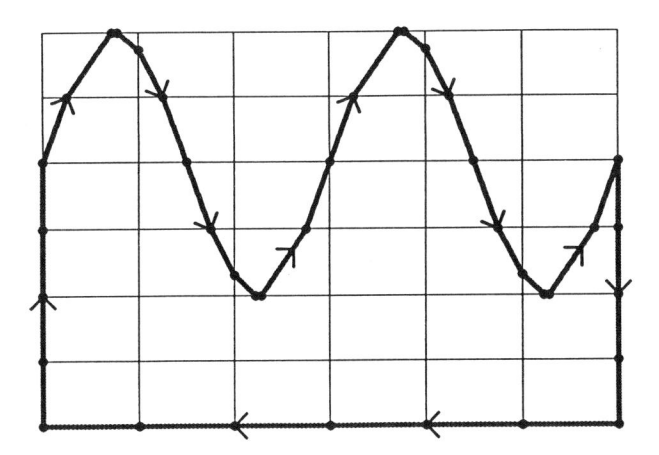

Figure 6: Meshed curve.

Computational information (such as the solution) is kept only at grid points, the points of intersection of grid lines. The division of grid blocks (the area bounded by neighboring grid lines) into many smaller subblocks is not computationally useful unless information can be associated with the subblocks, which would effectively yield a refined grid. In other words, if resolution finer than a grid block is required, then the grid should be refined. (At present, this must be done manually.) For this reason, multiple crossings of any grid block by a curve are eliminated in an operation called **pruning**, which we now describe.

In the following discussion, it will be useful to label as block (i,j) that grid block which encompasses the area between x_i and x_{i+1} and between y_j and y_{j+1}. To prune a meshed curve, we move along the meshed curve one chord (two consecutive curve points) at a time. If the chord crosses grid block (i,j), then we increment a counter $box(i,j)$. If $box(i,j)$ exceeds one, we move backwards along the curve, removing a point at a time from the curve structure, until $box(i,j) = 1$. Figure 7 shows the curve of Figure 6 after pruning.

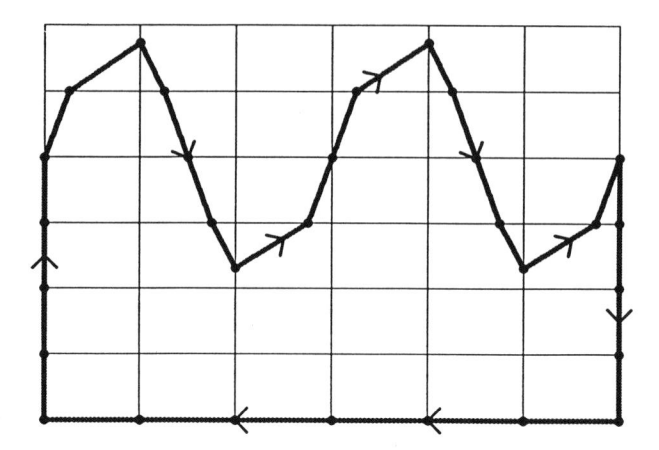

Figure 7: Pruned curve.

In keeping with the philosophy that information is to be kept only at grid points, the pruned curve is finally replaced with the closest possible match given by a sequence of grid points. To accomplish this, the "adjust" algorithm described above is applied with $2 \times smallx$ set to half the grid spacing (and similarly for $2 \times smally$). The resulting adjusted curve may contain redundant sequences of curve points, since several points may be adjusted to the same destination. Worse yet, the adjustment process can yield "kinks" if two non-consecutive curve points are adjusted to the same grid point while intervening curve points are not. The curve must therefore be "cleaned" by identifying and removing kinks and redundant points.

The cleaned curve may contain "gaps," e.g., consecutive points separated by more than one grid line. The cleaned curve is therefore re-meshed to yield the final result, the **projected** curve. The projected curve is continuous in the sense that consecutive curve points are separated by at most dx_i and dy_j. Figure 8 shows the curve of Figure 7 after projecting.

At first glance, the projected curve of Figure 8 may seem a poor representation of the boundary curve in Figure 5. However, this is merely an indication that the grid spacing may be inappropriate given the characteristics of the curve. Figure 9 shows the projection of the same curve onto a finer grid.

Projecting of boundary curves simplifies the algorithm described above, by which regions are defined by boundary curves. In addition, after projecting boundary curves, the smallest division of area is one half grid block. This information is used when applying the box method to points on the boundary of a region.

The experience of Glimm and McBryan was that projecting curves in this manner numerically stabilizes an interface, making the method inappropriate for simulations of interface instabilities, such as in gas dynamics. The slow oxide growth modeled here does not exhibit instabilities experimentally, however, and the projection method is quite appropriate.

To avoid accumulation of the positional error associated with projection, the projected curve is never propagated. Instead, at each timestep, the exact boundary curve is calculated and projected. The process of moving a boundary curve itself is thus straightforward, but after the movement, algorithms of greater complexity are required to deal with the consequences of the move, as we now describe.

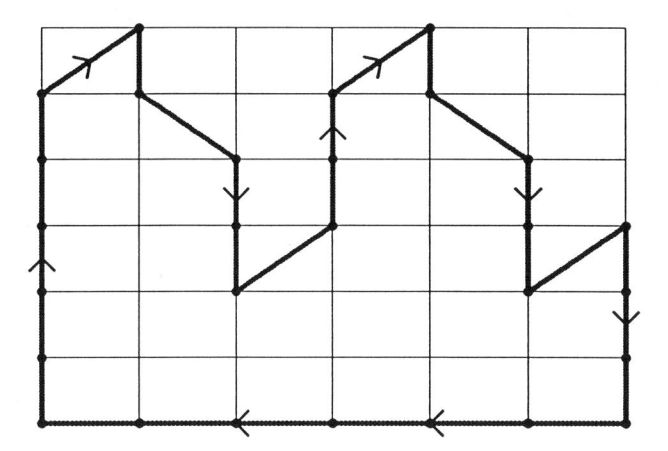

Figure 8: Projected curve.

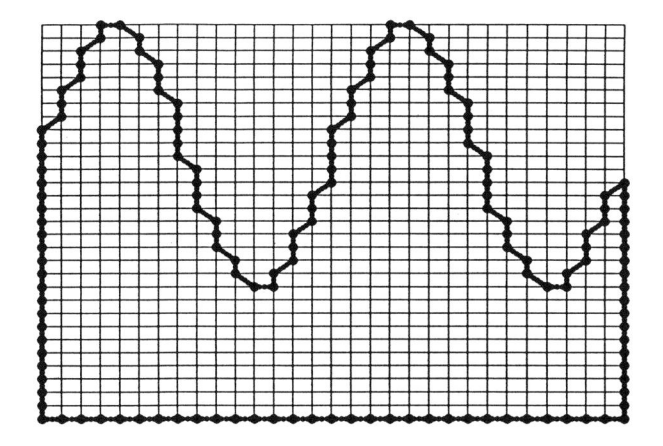

Figure 9: Projected curve on finer grid.

3.2. Transition regions. Physically, the moving boundary represents a growing oxide. Calculations are not performed on the oxide region. To avoid re-tallying the entire grid at each timestep, only the areas between the old and new boundary curves are reassigned after moving the boundary. These areas are called the **transition regions**. Figure 10 shows the transition regions (labeled I–IV) formed when the boundary curve of Figure 8 is moved to the position marked "new."

Boundary curves are always set up to encircle the regions they define in a clockwise manner. This convention is used to ensure a consistent definition of the concepts of moving "forward" or "backward" along a curve, and it is also employed when discretizing via the box method, to decide locally which octants are in the region.

The algorithm used to define each transition region is as follows:

Algorithm to define a single transition region.

- Move forward along the old boundary until a point is reached which is not in the new boundary.
- Move backward along the old boundary one point. Call this point A. Copy point A into the transition boundary as its first point.
- Move forward along the old boundary, copying each point into the transition boundary, until another point is reached which is in the new boundary. Call this point B. Copy point B into the transition boundary.
- Move backward along the new boundary, copying each point into the transition boundary, until point A is reached. Copy point A into the transition boundary as its last point.

As can be seen in Figure 10, a single boundary movement can result in many transition regions. Therefore, the transition region algorithm must be applied again and again, starting from point B, etc., until the last point in the old boundary is reached. Surprisingly, this still does not yield all the transition regions. The remainder must be found by reversing the roles of new and old above; that is, searching for points in the new boundary which are not in the old boundary. Transition region II in Figure 10 is of this type.

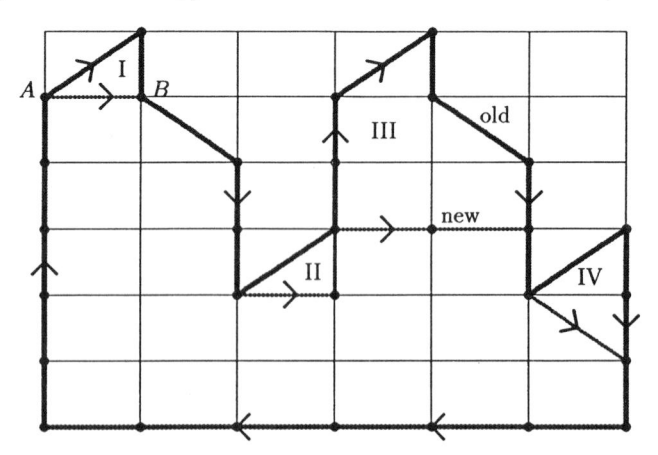

Figure 10: Transition regions between old and new boundary curves.

Points in the transition region can be identified by knowledge of the boundary. All these points must now be flagged as lying in the oxide region. Since points in the transition region were formerly in silicon, they are already flagged as lying in the silicon region as well. Points in the interior of the transition region must have their flags changed to indicate removal from the silicon region. Points on the new boundary are flagged as lying in both silicon and oxide regions.

Finally, the total amount of dopant in each transition region is calculated and added to the running total of dopant in the oxide. In this way, checks for conservation of total dopant can be implemented without recounting the oxide total; in fact, information about points in the oxide may in principle be discarded. Given homogeneous Neumann boundary conditions, the numerical methods employed here for the discretized problem (i.e., TRBDF2 timestepping with box method spatial discretization) conserve total dopant exactly in exact arithmetic. (The same boxes must be used as are employed in the box method; dopant per box is the concentration at the one internal grid point times box area.) Thus, conservation checks are useful rapid indicators of both software implementation error and rounding error due to finite machine precision.

3.3. Dynamic nature of curve structures. As curves are meshed, pruned, moved, etc., the number of points in the curve, and consequently the amount of storage required to represent the curve, may increase or diminish. In order to handle curves efficiently, dynamic allocation subroutines are used. These increase or decrease the storage allocated to a given curve structure. On the other hand, most curves contain many points, so that inefficiencies would result if the allocation routines were called each time a point is added or deleted. Therefore, storage for curves is increased and decreased in blocks which can hold many points.

4. Numerical experiments.

The numerical experiments described here have been summarized in the form of tables and graphs below. In the tables, in the column labeled "Case," M is a medium (40×20) grid and F is a fine (80×40) grid; SB refers to a stationary boundary and MB to a moving boundary; the final digit is the number of dopant species. The column labeled "MN" is the total number of unknowns. (M is the number of spatial points and N is the number of dopant species.)

The implant (initial data) is the following Gaussian, which yields a total dose of approximately 10.5×10^{20} atoms/cm per species:

$$C = e^{-30(\mathcal{X}^2 + \mathcal{Y}^2)} \times 10^{20} \text{ atoms/cm}^3,$$

$$\mathcal{X} = \frac{x - \frac{1}{2}(x_{\max} + x_{\min})}{x_{\max} - x_{\min}},$$

$$\mathcal{Y} = \frac{y - y_{\min}}{y_{\max} - y_{\min}}.$$

In cases of more than one species, the initial profiles are the same except that the concentration of the second species is 0.9 times that of the first. In all cases, the simulated annealing time is $T = 30$ min.

For the single species cases, the diffusivity was modeled as $D = aC + b$, with $a = 5 \times 10^{-33} \text{cm}^5/\text{sec}$ and $b = 0.1 \times 10^{-13} \text{cm}^2/\text{sec}$. For the dual species cases, the diffusivity matrix was

$$(17) \qquad D = \begin{pmatrix} aC_1 + b & -dC_1 \\ -dC_2 & aC_2 + b \end{pmatrix},$$

with $d = 0.05 \times 10^{-33} \text{cm}^5/\text{sec}$. The diagonal blocks in D crudely approximate the actual diffusivity matrix blocks for boron (C_1) and phosphorus (C_2) at 1000 C when the two concentrations are roughly equal. The off-diagonal blocks are greatly simplified from the actual cross terms, which decrease rapidly as the concentrations of the two species begin to differ significantly [9,10].

4.1. Timestep selection. As discussed in Section 2, the timestep size is adjusted according to the most recent estimate of the local truncation error τ^{n+1}, which can be obtained by either a divided difference formula or the TR/TR method. A comparison of the relative performance of these two methods is shown in Table 1, and condensed into scattergram form in Figure 11.

Table 1: Comparison of TR/TR and divided difference estimators. CPU times were measured on a Sun 4/280.

Case	MN	TR/TR CPU sec.	TR/TR time-steps	DD CPU sec.	DD time-steps	$\Delta t_{TR/TR}/\Delta t_{DD}$ min	$\Delta t_{TR/TR}/\Delta t_{DD}$ max
M-SB-1	800	108	12	89	12	1.00	1.09
M-MB-1	800	129	13	115	15	1.04	1.86
M-SB-2	1600	515	11	454	12	1.04	1.08
M-MB-2	1600	615	13	571	15	1.04	1.24
F-SB-1	3200	721	11	644	12	1.05	1.08
F-MB-1	3200	889	14	948	19	1.05	2.26
F-SB-2	6400	5558	11	4846	12	1.05	1.08
F-MB-2	6400	5458	13	6635	19	1.05	2.26

For stationary boundary problems, the divided difference error estimate gives superior performance. For large moving boundary cases, the TR/TR estimate is preferable, for the following reason. At the beginning of a timestep the boundary is moved to a place where ∇C was formerly nonzero. In calculating F^n (at the beginning of the timestep), we force $\hat{n} \cdot \nabla C = 0$ on the boundary; but only a short distance away from the boundary, the initial gradient of C (which was inherited from the preceding timestep) may be fairly steep. Since F is calculated from second spatial derivatives, we may expect the initial F^n to contain some error associated with this effect near the boundary. The other terms in Eq. (14), $F^{n+\gamma}$ and F^{n+1}, are calculated from C's that have undergone diffusion since the boundary was moved, so they do not contain this error. Since τ^{n+1} is calculated from *differences* in these values, a given relative error in F^n will induce a relatively larger error in τ^{n+1} and, hence, in the calculated timestep size.

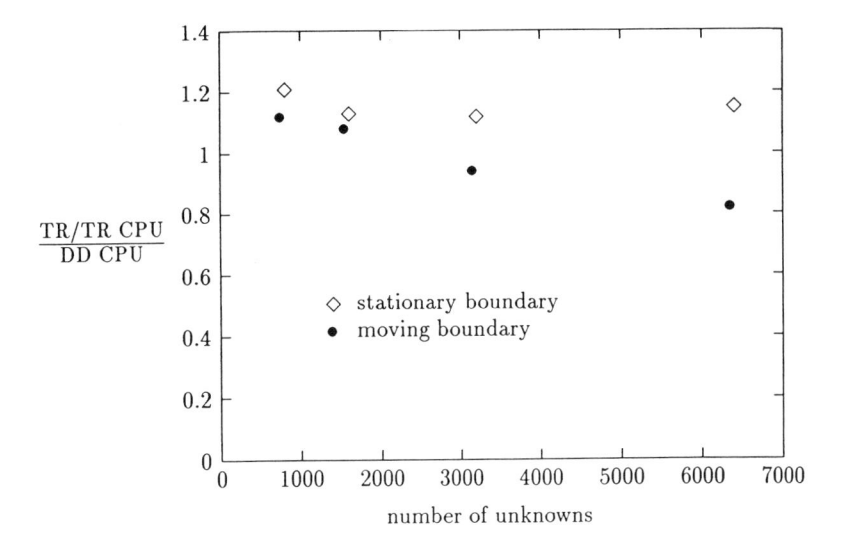

Figure 11: Relative CPU usage for TR/TR and divided difference timestep adjusters.

4.2. Re-using Jacobians. If the Jacobian is changing slowly enough, there is a potential for savings in execution time by re-using old factorizations, that is, by employing Newton-Richardson iteration as opposed to Newton iteration. On the other hand, since the reused Jacobian will no longer generate updates along the true Newton direction, the number of iterations required for convergence may increase. Software algorithms implementing Newton-Richardson iteration must therefore include some test for slow convergence, at which point the matrix is re-factored. Since the assumption of a slowly changing Jacobian is then suspect, a flag may be set to re-factor on subsequent iterations as well [11,3].

Our code includes three such tests. Ordinarily, we factor the matrix only at the beginning of each composite TRBDF2 timestep. However, the matrix is re-factored in any of the following circumstances:

(a) If MAX-NEWTONS are reached without convergence, the code switches to standard Newton iteration for the duration of the timestep. (MAX-NEWTONS has been set at 9.)

(b) If the norm of the residual exceeds MAX-RATIO times the norm of the residual at the previous iteration, the Jacobian is re-factored. (MAX-RATIO has been set at 0.1, which implies that re-factorization occurs if the Newton-Richardson method is producing less than one decimal digit per iteration.)

(c) If the damping factor λ_{k+1} is less than 1.0, the Jacobian is re-factored. (The damping factor is calculated by the formula of Bank and Rose [3].)

The performance of this Newton-Richardson method was compared experimentally with that of ordinary Newton iteration. The results are summarized in Table 2. The Newton-Richardson method usually gives a modest performance improvement over the Newton method, and occasionally only a slight degradation, so that Newton-Richardson may be considered as the method of choice for most nonlinear diffusion problems.

Table 2: Comparison of Newton-Richardson and Newton methods.

Case	MN	time-steps	Newton-Richardson CPU sec.	Newton-Richardson total iterations	Newton CPU sec.	Newton total iterations
M-SB-1	800	12	87	51	89	46
M-MB-1	800	15	103	58	115	57
M-SB-2	1600	12	466	49	454	45
M-MB-2	1600	15	521	58	571	57
F-SB-1	3200	12	670	51	644	46
F-MB-1	3200	19	926	74	948	69
F-SB-2	6400	12	4793	59	4846	58
F-MB-2	6400	19	6007	74	6635	74

5. Conclusion.

We have demonstrated an efficient set of algorithms appropriate for modeling stable interfaces in two spatial dimensions, and we have applied these algorithms to the solution of a set of nonlinear diffusion equations on a region with a moving boundary, from the field of semiconductor process modeling. We have shown that for problems with a stationary boundary, a divided difference error estimator gives optimal performance, while a TR/TR scheme is preferable with a moving boundary. We have also demonstrated that in most nonlinear diffusion problems, Newton-Richardson iteration yields a modest performance improvement over Newton iteration.

REFERENCES

[1] R.B. FAIR, C.L. GARDNER, M.J. JOHNSON, S.W. KENKEL, D.J. ROSE, J.E. ROSE, AND R. SUBRAHMANYAN, *Two dimensional process simulation using verified phenomenological models*, IEEE Transactions on Computer-Aided Design of Integrated Circuits and Systems, vol. 10 (1991), pp. 643–651.

[2] R.E. BANK, W.M. COUGHRAN, W. FICHTNER, E.H. GROSSE, D.J. ROSE, AND R.K. SMITH, *Transient simulation of silicon devices and circuits*, IEEE Transactions on Computer-Aided Design, vol. CAD-4 (1985), pp. 436–451.

[3] R.E. BANK AND D.J. ROSE, *Global approximate Newton methods*, Numerische Mathematik, vol. 37 (1981), pp. 279–295.

[4] R.E. BANK, D.J. ROSE, AND W. FICHTNER, *Numerical methods for semiconductor device simulation*, SIAM Journal on Scientific and Statistical Computing, vol. 4 (1983), pp. 416–435.

[5] R.S. VARGA, *Matrix Iterative Analysis*, Prentice-Hall, 1962.

[6] H.R. YEAGER AND R.W. DUTTON, *An approach to solving multiparticle diffusion exhibiting nonlinear stiff coupling*, IEEE Transactions on Electron Devices, vol. ED-32 (1985), pp. 1964–1976.

[7] I.L. CHERN, J. GLIMM, O. MCBRYAN, B. PLOHR, AND S. YANIV, *Front tracking for gas dynamics*, Journal of Computational Physics, vol. 62 (1986), pp. 83–110.

[8] D.H. SELIM, *Tensor product grid implementation for PREDICT2*, Master's thesis, Duke University, 1990.

[9] Microelectronics Center of North Carolina, Research Triangle Park, NC, *PREDICT Users Manual*, 1986.

[10] M.J. JOHNSON, *Numerical Methods for Semiconductor Process Simulation in Two Spatial Dimensions: a Nonlinear Diffusion Problem with a Free Boundary*, PhD thesis, Duke University, 1991.

[11] W.M. COUGHRAN, E.H. GROSSE, AND D.J. ROSE, *Aspects of computational circuit analysis*, in VLSI CAD Tools and Applications (W. Fichtner and M. Morf, eds.), pp. 105–127, Kluwer Publishers, Boston, 1986.

ASYMPTOTIC ANALYSIS OF A MODEL FOR THE DIFFUSION OF DOPANT–DEFECT PAIRS

J.R. KING*

Abstract. Asymptotic methods are applied to a model describing the diffusion through silicon of a dopant which pairs with both vacancies and self-interstitials. Several different asymptotic limits are discussed for problems in both one and higher dimensions.

1. Introduction and model. The purpose of this paper is to summarise some results of an asymptotic analysis of models for the diffusion of a dopant in silicon mediated by mobile dopant-point defect pairs. Such models are now widely accepted as providing accurate descriptions of the redistribution of many impurities in semiconductors; see, for example, Fahey et al. [1]. The defects involved are the vacancy (V) and the self-interstitial (I), each of which may pair with a dopant atom (d) to produce a pair (dV or dI). For simplicity we shall follow Richardson and Mulvaney [10] and Morehead and Lever [9] in treating all species as electrically neutral. Six bimolecular reactions between the various species are then possible, as follows.

		Forward reaction rate	Reverse reaction rate
(1)	$V + I \rightleftharpoons 0$	$F_1 = K_1 c_V c_I$	$R_1 = K_1 c_V^* c_I^*$
(2)	$d + V \rightleftharpoons dV$	$F_2 = K_2 \omega_V c_d c_V$	$R_2 = K_2 c_V^* c_{dV}$
(3)	$d + I \rightleftharpoons dI$	$F_3 = K_3 \omega_I c_d c_I$	$R_3 = K_3 c_I^* c_{dI}$
(4)	$dV + I \rightleftharpoons d$	$F_4 = K_4 c_{dV} c_I$	$R_4 = K_4 \omega_V c_I^* c_d$
(5)	$dI + V \rightleftharpoons d$	$F_5 = K_5 c_{dI} c_V$	$R_5 = K_5 \omega_I c_V^* c_d$
(6)	$dV + dI \rightleftharpoons 2d$	$F_6 = K_6 c_{dV} c_{dI}$	$R_6 = K_6 \omega_V \omega_I c_d^2$

Here c with the appropriate suffix denotes the concentration of a given species, c_V^* and c_I^* are equilibrium concentrations of vacancies and interstitials respectively, and $K_1 - K_6$ and ω_V and ω_I are constants. We note that the K's all have the same dimensions and that ω_V and ω_I are dimensionless. Relationships between various reaction coefficients have been derived by exploiting the principle of detailed balance (see, for example, [7]) which implies that under equilibrium conditions each reaction must individually be in balance. Thus in equilibrium

$$c_V^* c_I^* = c_V c_I, \qquad c_{dV} = \omega_V c_d c_V \big/ c_V^*, \qquad c_{dI} = \omega_I c_d c_I \big/ c_I^*,$$

---

*Department of Theoretical Mechanics, University of Nottingham NG7 2RD, England

which implies in particular that if, in addition, defect concentrations individually take their equilibrium values then we have

$$c_{dV} = \omega_V c_d, \qquad c_{dI} = \omega_I c_d.$$

The corresponding system of reaction-diffusion equations is

(1.1) $\quad \dfrac{\partial c_d}{\partial t} = R_2 - F_2 + R_3 - F_3 + F_4 - R_4 + F_5 - R_5 + 2F_6 - 2R_6,$

(1.2) $\quad \dfrac{\partial c_{dV}}{\partial t} = D_{dV}\dfrac{\partial^2 c_{dV}}{\partial x^2} + F_2 - R_2 + R_4 - F_4 + R_6 - F_6,$

(1.3) $\quad \dfrac{\partial c_{dI}}{\partial t} = D_{dI}\dfrac{\partial^2 c_{dI}}{\partial x^2} + F_3 - R_3 + R_5 - F_5 + R_6 - F_6,$

(1.4) $\quad \dfrac{\partial c_V}{\partial t} = D_V\dfrac{\partial^2 c_V}{\partial x^2} + R_1 - F_1 + R_2 - F_2 + R_5 - F_5,$

(1.5) $\quad \dfrac{\partial c_I}{\partial t} = D_I\dfrac{\partial^2 c_I}{\partial x^2} + R_1 - F_1 + R_3 - F_3 + R_4 - F_4,$

where the D's are constant diffusivities and it is assumed that the dopant is unable to diffuse in its unpaired state. Before proceeding further we make the simplifying assumption that K_2 and K_3 are sufficiently large that the relations

(1.6) $\qquad c_{dV} = \omega_V c_d c_V / c_V^*, \qquad c_{dI} = \omega_I c_d c_I / c_I^*$

may be assumed valid. The governing system is then made up of the algebraic equations (1.6) together with the following:

(1.7) $\quad \dfrac{\partial}{\partial t}\left(c_d + c_{dV} + c_{dI}\right) = \dfrac{\partial^2}{\partial x^2}\left(D_{dV}c_{dV} + D_{dI}c_{dI}\right),$

(1.8) $\quad \dfrac{\partial}{\partial t}\left(c_V + c_{dV}\right) = \dfrac{\partial^2}{\partial x^2}\left(D_V c_V + D_{dV} c_{dV}\right) + R_1 - F_1 + R_4 - F_4 + R_5 - F_5 + R_6 - F_6,$

(1.9) $\quad \dfrac{\partial}{\partial t}\left(c_I + c_{dI}\right) = \dfrac{\partial^2}{\partial x^2}\left(D_I c_I + D_{dI} c_{dI}\right) + R_1 - F_1 + R_4 - F_4 + R_5 - F_5 + R_6 - F_6,$

which are obtained by taking suitable combinations of (1.1) – (1.5). We note that conditions such as (1.6) must hold in order for the governing equation to be the linear diffusion equation

$$\dfrac{\partial c_d}{\partial t} = D_i \dfrac{\partial^2 c_d}{\partial x^2}$$

when c_d is small everywhere and $c_V \sim c_V^*$, $c_I \sim c_I^*$ everywhere, in this case the intrinsic diffusivity D_i being given by

$$D_i = \left(D_{dV}\omega_V + D_{dI}\omega_I\right)\big/\left(1 + \omega_V + \omega_I\right).$$

We note, however, that (1.7) – (1.9) is a lower order system than (1.1) – (1.5), and if the boundary conditions on (1.1) – (1.5) are not consistent with (1.6) then boundary layers will occur in which (1.6) is not valid. Additional timescales are also necessary to describe initial transients when (1.6) does not hold at $t = 0$.

Using (1.6) we may rewrite (1.8) and (1.9) as

(1.10) $$\frac{\partial}{\partial t}\left(c_V + c_{dV}\right) = \frac{\partial^2}{\partial x^2}\left(D_V c_V + D_{dV} c_{dV}\right)$$
$$+ \left(K_1 + \left(K_4 \frac{\omega_V}{c_V^*} + K_5 \frac{\omega_I}{c_I^*}\right)c_d + K_6 \frac{\omega_V \omega_I}{c_V^* c_I^*} c_d^2\right)\left(c_V^* c_I^* - c_V c_I\right),$$

(1.11) $$\frac{\partial}{\partial t}\left(c_I + c_{dI}\right) = \frac{\partial^2}{\partial x^2}\left(D_I c_I + D_{dI} c_{dI}\right)$$
$$+ \left(K_1 + \left(K_4 \frac{\omega_V}{c_V^*} + K_5 \frac{\omega_I}{c_I^*}\right)c_d + K_6 \frac{\omega_V \omega_I}{c_V^* c_I^*} c_d^2\right)\left(c_V^* c_I^* - c_V c_I\right),$$

from which the enhancement to the defect generation-recombination rate resulting from the presence of dopant is apparent.

2. Non-dimensionalization and preliminary asymptotics. We now non-dimensionalize (1.6), (1.7), (1.10) and (1.11) by writing

$$c_d = c_d^* u, \quad c_V = c_V^* V, \quad c_I = c_I^* I, \quad c_{dV} = \omega_V c_d^* P_V, \quad c_{dI} = \omega_I c_I^* P_I,$$
$$t = T\bar{t}, \quad x = ((D_{dV}\omega_V + D_d\omega_I)T)^{\frac{1}{2}}\bar{x},$$

where c_d^* is a representative dopant concentration and T is a representative timescale; we note that V and I now denote dimensionless concentrations. We then obtain (dropping overbars)

(2.1) $P_V = uV, \quad P_I = uI,$

(2.2) $$\frac{\partial}{\partial t}\left((1 + \omega_V V + \omega_I I)u\right) = \frac{\partial^2}{\partial x^2}\left((f_V V + f_I I)u\right),$$

(2.3) $$\frac{\partial}{\partial t}\left((r_V + \omega_V u)V\right) = \frac{\partial^2}{\partial x^2}\left((g_V r_V + f_V u)V\right) + \left(\kappa_0 r + \kappa_1 u + \frac{\kappa_2}{r} u^2\right)(1 - IV)$$

(2.4) $$\frac{\partial}{\partial t}\left((r_I + \omega_I u)I\right) = \frac{\partial^2}{\partial x^2}\left((g_I r_I + f_I u)I\right) + \left(\kappa_0 r + \kappa_1 u + \frac{\kappa_2}{r} u^2\right)(1 - IV),$$

where we have dimensionless constants

$$r_V = c_V^*/c_d^*, \quad r_I = c_I^*/c_d^*, \quad r = (r_V r_I)^{\frac{1}{2}},$$
$$f_V = D_{dV}\omega_V/(D_{dV}\omega_V + D_{dI}\omega_I), \quad f_I = D_{dI}\omega_I/(D_{dV}\omega_V + D_{dI}\omega_I),$$
$$g_V = D_V/(D_{dV}\omega_V + D_{dI}\omega_I), \quad g_I = D_I/(D_{dV}\omega_V + D_{dI}\omega_I),$$
$$\kappa_0 = K_1\left(c_V^* c_I^*\right)^{\frac{1}{2}} T, \quad \kappa_1 = \left(K_4\omega_V c_I^* + K_5\omega_I c_V^*\right)T, \quad \kappa_2 = K_6\left(c_V^* c_I^*\right)^{\frac{1}{2}} T.$$

We note the relationship
$$f_V + f_I = 1,$$
and that r_V, r_I and r are the only parameters which depend on c_d^*. The notation used in (2.1) – (2.4) is somewhat different from that used earlier [4] in discussing a special case of this model.

We now note two sets of boundary and initial conditions applicable to (2.2) – (2.4), as follows.

(a) Surface source

(2.5)
$$\begin{cases} \text{at } x = 0 & u = u_s, \quad V = 1,\ I = 1, \\ \text{as } x \to +\infty & u \to 0, \quad V \to 1,\ I = 1, \\ \text{at } t = 0 & u = 0, \quad V = 1,\ I = 1, \end{cases}$$

where u_s is a prescribed constant.

(b) Implanted dopant

(2.6)
$$\begin{cases} \text{at } x = 0 & \dfrac{\partial}{\partial x}\left((f_V V + f_I I)u\right) = 0, \quad V = 1,\ I = 1, \\ \text{as } x \to +\infty & u \to 0, \quad V \to 1,\ I = 1, \\ \text{at } t = 0 & u = U(x), \quad V = 1,\ I = 1, \end{cases}$$

where $U(x)$ is a prescribed function with
$$Q = \int_0^\infty U(x)\,dx$$

finite. The initial conditions on V and I neglect implantation damage effects, but these initial conditions will in any case play little role in the subsequent analysis. It is convenient to write
$$\text{at } x = 0 \quad u = u_s(t);$$
in this case u_s must be determined as part of the solution.

Based on the parameters used by Morehead and Lever [9] we have the following approximate orders of magnitude (taking c_d^* to be the surface dopant concentration and T to be the duration of the diffusion):

$$r_V,\ r_I,\ r \sim 10^{-9},$$
$$\omega_V,\ \omega_I \sim 10^{-3},$$
$$g_V,\ g_I \sim 10^6.$$

Morehead and Lever [9] do not include terms corresponding to κ_0 or κ_2 and they assume that κ_1 is sufficiently large that the generation recombination term dominates giving

$$IV \sim 1.$$

The parameters used by Richardson and Mulvaney [10] also imply that r_V, $r_I r$, ω_V and ω_I are small and that g_V and g_I are large.

We shall always therefore consider the limits

$$r_V, r_I, r, \omega_V, \omega_I \to 0,$$

and in much of what follows we shall concentrate on the case in which

$$g_V r_V, \quad g_I r_I = O(1)$$

and we assume that

$$r_I = O(r_V), \quad g_I = O(g_V), \quad t = O(1).$$

In this case the solution has a two region asymptotic structure made up of an inner region ($x = O(1)$) on the length scale of dopant diffusion and an outer region $\left(x = O(g^{\frac{1}{2}}) \text{ where } g = (g_V g_I)^{\frac{1}{2}}\right)$ on the much longer length scale of defect diffusion.

In this limit, equations (2.3) and (2.4) imply that for $x = O(1)$ we have at leading order

$$\frac{\partial^2}{\partial x^2}\left((g_V r_V + f_V u)V\right) = \frac{\partial^2}{\partial x^2}\left((g_I r_I + f_I u)I\right).$$

Matching into the outer region then implies that

(2.7) $$\frac{\partial}{\partial x}\left((g_V r_V + f_V u)V\right) = \frac{\partial}{\partial x}\left((g_I r_I + f_I u)I\right),$$

which corresponds to the well-established assumption of flux balance (see, for example, [11], [8]).

In each of cases (a) and (b) above, equation (2.7) implies that

(2.8) $$(g_V r_V + f_V u)V - (g_I r_I + f_I u)I = g_V r_V - g_I r_I + (f_V - f_I)u_s,$$

and the leading order problem in $x = O(1)$ is governed by (2.8) together with

(2.9) $$\frac{\partial u}{\partial t} = \frac{\partial^2}{\partial x^2}\left((f_V V + f_I I)u\right),$$

(2.10) $$\frac{\partial^2}{\partial x^2}\left((g_V r_V + f_V u)V\right) + \left(\kappa_0 r + \kappa_1 u + \frac{\kappa_2}{r}u^2\right)(1 - IV) = 0,$$

where we have retained all the generation-recombination terms.

In the outer region we write $x = g^{\frac{1}{2}}y$, and for $y = O(1)$, $g \gg 1$ the dopant concentration u is exponentially small with respect to g and the dominant balance is given by

$$\frac{\partial V}{\partial t} = \gamma \frac{\partial^2 V}{\partial y^2} + \frac{\kappa_0}{\rho}(1 - IV), \tag{2.11}$$

$$\frac{\partial I}{\partial t} = \frac{1}{\gamma}\frac{\partial^2 I}{\partial y^2} + \kappa_0 \rho (1 - IV), \tag{2.12}$$

where $\gamma = (g_V/g_I)^{\frac{1}{2}}$ and $\rho = (r_V/r_I)^{\frac{1}{2}}$. Matching the two regions together implies that the conditions

$$\text{as } x \to +\infty \quad u \to 0, \quad \frac{\partial V}{\partial x} \to 0 \tag{2.13}$$

hold on (2.9) and (2.10), and defining V_∞ and I_∞ by

$$\text{as } x \to +\infty \quad V(x,t) \to V_\infty(t), \quad I(x,t) \to I_\infty(t),$$

where V and I denote leading order solutions to the inner problem, then (2.11) and (2.12) are governed by

$$\begin{aligned}
\text{at } y = 0 \quad & V = V_\infty(t), \quad I = I_\infty(t), \\
\text{as } y \to +\infty \quad & V \to 1, \quad I \to 1, \\
\text{at } t = 0 \quad & V = 1, \quad I = 1.
\end{aligned} \tag{2.14}$$

In order to make further analytical progress we now consider limiting processes involving the generation-recombination terms, the two possible limits being discussed in the next two sections.

3. Negligible generation-recombination.

3.1 Intermediate dopant concentrations $r = O(1/g)$. The precise balance of terms in (2.2) – (2.4) depends on the maximum dopant concentration present, and there are two cases we need to discuss, namely an intermediate concentration case in which $r = O(1/g)$ and a high concentration case with $r = O(\omega/g)$, where $\omega = (\omega_V \omega_I)^{\frac{1}{2}}$ and we assume that $\omega_I = O(\omega_V)$. Here the value of r is determined by defining c_d^* to be the surface concentration of dopant in the surface source case and to be the maximum value of c_d in the initial profile for the ion-implanted case.

We start by assuming that $r = O(1/g)$, which was the case considered in the previous section and which led to the flux-balance condition (2.7). We now assume in addition that

$$\kappa_0 r, \quad \kappa_1, \quad \kappa_2/r \ll 1,$$

so that for $x = O(1)$ the generation-recombination terms in (2.3) and (2.4) may be neglected. This implies that at leading order

$$\frac{\partial^2}{\partial x^2}\left((g_V r_V + f_V u)V\right) = \frac{\partial^2}{\partial x^2}\left((g_I r_I + f_I u)I\right) = 0$$

so that

(3.1) $$V = \left(g_V r_V + f_V u_s\right)/\left(g_V r_V + f_V u\right),$$

(3.2) $$I = \left(g_I r_I + f_I u_s\right)/\left(g_I r_I + f_I u\right),$$

and then (2.9) becomes

(3.3) $$\frac{\partial u}{\partial t} = \frac{\partial}{\partial x}\left(D_1(u, u_s)\frac{\partial u}{\partial x}\right)$$

where

(3.4) $$D_1(u, u_s) = f_V \frac{(1 + s_V u_s)}{(1 + s_V u)^2} + f_I \frac{(1 + s_I u_s)}{(1 + s_I u)^2}$$

with

(3.5) $$s_V = f_V/g_V r_V, \qquad s_I = f_I/g_I r_I.$$

We note that the contributions of vacancy and interstitial mechanisms to the effective diffusivity D_1 are in this case additive. It follows from (3.1) and (3.2) that in (2.14) we have

(3.6) $$V_\infty = 1 + s_V u_s, \qquad I_\infty = 1 + s_I u_s,$$

so that s_V and s_I provide measures of the degree of point defect supersaturation.

The diffusivity (3.4) exhibits a surface concentration enhancement, the tail diffusivity giving given by

$$D_1(0, u_s) = 1 + (f_V s_V + f_I s_I)u_s.$$

The corresponding profile for u will not, however, exhibit at high concentrations a plateau of the kind that is observed for phosphorus diffusion in silicon (see [3]). This follows because D_1 decreases with increasing u.

We note from (3.6) that if $gr \ll 1$ then the defect supersaturation is very large. If it is sufficiently large then the number of pairs becomes comparable to the number of unpaired dopant atoms and the dominant balance of terms changes. This occurs when $gr = O(\omega)$.

3.2 Very high dopant concentrations $r = O(\omega/g)$. The governing equations are (2.2) together with

$$\frac{\partial}{\partial t}\left((r_V + \omega_V u)V\right) = \frac{\partial^2}{\partial x^2}\left((g_V + f_V u)V\right),$$

and

$$\frac{\partial}{\partial t}\left((r_I + \omega_I u)I\right) = \frac{\partial^2}{\partial x^2}\left((g_I + f_I u)I\right).$$

We restrict attention here to the surface source case (2.5), in which case the solution is self-similar with $u \equiv u(x/t^{\frac{1}{2}})$, $V \equiv V(x/t^{\frac{1}{2}})$, $I \equiv I(x/t^{\frac{1}{2}})$. Guided by (3.6), for $r = O(\omega/g)$ we introduce the rescalings

(3.7) $\qquad V = \widehat{V}/\omega, \qquad I = \widehat{I}/\omega, \qquad u = \omega\widehat{u}, \qquad x = \widehat{x}/\omega^{\frac{1}{2}},$

and write

$$h_V = g_V r_V\big/\omega, \qquad h_I = g_I r_I\big/\omega, \qquad \nu = \left(\omega_V\big/\omega_I\right)^{\frac{1}{2}},$$

to give the following leading order balance in $\widehat{x} = O(1)$:

(3.8)
$$\begin{cases} \dfrac{\partial \widehat{u}}{\partial t} = -\dfrac{\partial^2}{\partial \widehat{x}^2}\left(h_V \widehat{V} + h_I \widehat{I}\right), \\ \dfrac{\partial}{\partial t}\left(\nu \widehat{V} \widehat{u}\right) = \dfrac{\partial^2}{\partial \widehat{x}^2}\left((h_V + f_V \widehat{u})\widehat{V}\right), \\ \dfrac{\partial}{\partial t}\left(\widehat{I} \widehat{u}/\nu\right) = \dfrac{\partial^2}{\partial \widehat{x}^2}\left((h_I + f_I \widehat{u})\widehat{I}\right); \end{cases}$$

we note that flux-balance does not occur in this case.

It is clear from the rescalings (3.7) that since $\omega \ll 1$ the solution to (3.8) will not enable us to satisfy the conditions of (2.5) on $x = 0$. It turns out that there is a boundary layer in which we write

$$\widehat{x} = \omega \, \ln^{-\frac{1}{2}}(1/\omega) \, x^{\dagger}$$

and obtain (imposing conditions from (2.5))

(3.9) $\qquad V \sim u_s/u, \qquad I \sim u_s/u, \qquad u \sim u_s\Big/\left(1 + x^{\dagger}\left(u_s\big/(h_V + h_I)t\right)^{\frac{1}{2}}\right);$

the required solution to (3.8) satisfies

as $\widehat{x} \to 0^+ \quad \widehat{u} \sim \left((h_V + h_I)u_s t\right)^{\frac{1}{2}}\Big/\widehat{x} \, \ln^{\frac{1}{2}}(1/\widehat{x}), \qquad \widehat{V} \sim u_s/\widehat{u}, \qquad \widehat{I} \sim u_s/\widehat{u}.$

4. Dominant generation–recombination.

4.1 Intermediate dopant concentrations $r = O(1/g)$. We now consider the case in which at least one

$$\kappa_0 r \gg 1, \quad \kappa_1 \gg 1 \quad \text{and} \quad \kappa_2/r \gg 1$$

hold, so that for $u = O(1)$ equations (2.3) and (2.4) imply that at leading order

(4.1) $$IV = 1 .$$

For $r = O(1/g)$ equation (2.8) holds, and it then follows from (4.1) that

(4.2) $$V = \left\{ \left((g_I r_I - g_V r_V + (f_I - f_V) u_s)^2 + 4(g_I r_I + f_I u)(g_V r_V + f_V u) \right)^{\frac{1}{2}} \right.$$
$$\left. - (g_I r_I - g_V r_V + (f_I - f_V) u_s) \right\} / 2(g_V r_V + f_V u) ,$$

(4.3) $$I = \left\{ \left((g_I r_I - g_V r_V + (f_I - f_V) u_s)^2 + 4(g_I r_I + f_I u)(g_V r_V + f_V u) \right)^{\frac{1}{2}} \right.$$
$$\left. + (g_I r_I - g_V r_V + (f_I - f_V) u_s) \right\} / 2(g_I r_I + f_I u) .$$

Substituting into (2.9) now gives

(4.4) $$\frac{\partial u}{\partial t} = \frac{\partial}{\partial x} \left(D_2(u, u_s) \frac{\partial u}{\partial x} \right)$$

with
(4.5)
$$D_2(u, u_s) = \left((g_V r_V V + g_I r_I I)(f_V V + f_I I) + 4 f_V f_I u \right) / \left((g_V r_V V + g_I r_I I) + (f_V V + f_I I) u \right)$$

An explicit expression for the dependence of D_2 on u and u_s may be obtained by substituting for V and I using (4.2) and (4.3). Here we restrict further attention to some special cases. Firstly we note that if $f_V = f_I = \frac{1}{2}$ then (4.2) and (4.3) yield

$$V = 1 , \quad I = 1 ,$$

so that
$$\frac{\partial u}{\partial t} = \frac{\partial^2 u}{\partial x^2} .$$

Of more interest is the case in which $gr \ll 1$ when three different scales for u must be considered. We assume that $f_I > f_V$; if $f_V > f_I$ the roles of vacancies and interstitials are interchanged.

(i) $u = O(1)$ (high concentrations) (surface region).

For $g_V r_V$, $g_I r_I \ll 1$ we then obtain

$$V \sim \left\{\left((f_I - f_V)^2 u_s^2 + 4f_I f_V u^2\right)^{\frac{1}{2}} - (f_I - f_V)u_s\right\}\Big/ 2f_V u ,$$

$$I \sim \left\{\left((f_I - f_V)^2 u_s^2 + 4f_I f_V u^2\right)^{\frac{1}{2}} + (f_I - f_V)u_s\right\}\Big/ 2f_I u ,$$

and

(4.6) $\qquad D_2(u, u_s) \sim 4f_I f_V u \Big/ \left((f_I - f_V)^2 u_s^2 + 4f_I f_V u^2\right)^{\frac{1}{2}} .$

We note that in this case the terms $f_V V u$ and $f_I I u$ on the right-hand side of (2.9) make equal contributions, whatever the value of f_V/f_I (provided that $f_V, f_I \neq 0$)

(ii) $u = O\left((g_I r_I)^{\frac{1}{3}}\right)$ (intermediate concentrations) (kink region).

Then

$$V \sim f_I u \Big/ (f_I - f_V) u_s ,$$

$$I \sim (f_I - f_V) u_s \Big/ f_I u ,$$

and

(4.7) $\qquad D_2(u, u_s) \sim \dfrac{4f_I f_V u}{(f_I - f_V)u_s} + \dfrac{g_I r_I (f_I - f_V) u_s}{f_I u^2} .$

(iii) $u = O(g_I r_I)$ (low concentrations) (tail region).

We now have

$$V \sim (g_I r_I + f_I u) \Big/ (f_I - f_V) u_s ,$$

(4.8) $\qquad I \sim (f_I - f_V) u_s \Big/ (g_I r_I + f_I u),$

and

(4.9) $\qquad D_2(u, u_s) \sim g_I r_I f_I (f_I - f_V) u_s \Big/ (g_I r_I + f_I u)^2 .$

We note that which defect is supersaturated and which is undersaturated here depends only on which of f_I and f_V is the larger. We also note that there is again a surface concentration enhancement of the tail diffusivity with

$$D_2(0, u_s) \sim f_I(f_I - f_V) u_s \Big/ g_I r_I .$$

A uniformly valid approximation to D_2 is easily obtained in the form

(4.10) $\qquad D_2(u, u_s) \sim \dfrac{4f_I f_V u}{\left((f_1 - f_V)^2 u_s^2 + 4f_I f_V u^2\right)^{\frac{1}{2}}} + \dfrac{g_I r_I f_I (f_I - f_V) u_s}{(g_I r_I + f_I u)^2} .$

The three region structure given by (i), (ii) and (iii) is reminiscent of that used by Fair and Tsai [3] to describe phosphorus diffusion in silicon. The diffusivity is largest in the tail region (iii), drops to a minimum in the intermediate region (ii) and increases again in the high concentration region (i). It is evident from (4.6) that in order to obtain a plateau effect due to a sufficiently large diffusivity at high concentrations, we require that neither f_I nor f_V be too small. To obtain a significant tail diffusivity (4.9) we require that f_I and f_V not be too close in value.

We note from (4.8) that

$$\text{as} \quad u \to 0 \quad I \to (f_I - f_V)u_s \Big/ g_I r_I \; ,$$

so that the interstitial supersaturation becomes very large as $g_I r_I$ becomes small; if it is sufficiently large then the number of interstitial-dopant pairs can be comparable to the number of unpaired dopant atoms and a different balance of terms is again needed. We now discuss this high supersaturation case.

4.2 Very high dopant concentrations $r = O(\omega/g)$. The governing equations are (2.2), (4.1) and

$$\frac{\partial}{\partial t}\left((r_V + \omega_V u)V - (r_I + \omega_I u)I\right) = \frac{\partial^2}{\partial x^2}\left((g_V r_V + f_V u)V - (g_I r_I + f_I u)I\right) .$$

We again restrict attention to (2.5), in which case $u \equiv u(x/t^{\frac{1}{2}})$, $V \equiv V(x/t^{\frac{1}{2}})$ and $I \equiv I(x/t^{\frac{1}{2}})$ again hold. We again treat the case $f_I > f_V$. The asymptotic structure is complicated and we simply outline the various regions. We again write

$$g_I r_I = h_I \omega \; ,$$

with $h_I = O(1), \omega \ll 1$. The five regions describing the behaviour of the dopant are as follows.

(1) Surface region $u = O(1)$.

Writing

$$u = u_0(x,t) + o(1) \quad \text{as} \quad \omega \to 0$$

we have

$$(4.11) \qquad \frac{\partial u_0}{\partial t} = 4 f_I f_V \frac{\partial}{\partial x}\left(\frac{u_0}{\left((f_I - f_V)^2 u_s^2 + 4 f_I f_V u_0^2\right)^{\frac{1}{2}}} \frac{\partial u_0}{\partial x}\right)$$

(cf. (4.6)). Equation (4.11) is to be solved as a moving boundary problem subject to

$$(4.12) \qquad \begin{array}{lll} \text{at} & x = 0 & u_0 = u_s \; , \\ \text{at} & x = s_0(t) & u_0 = 0 \; , \quad \dfrac{\partial u_0}{\partial x} = -(f_I - f_V)u_s \dot{s}_0 \Big/ 4 f_I f_V \; , \\ \text{at} & t = 0 & u_0 = 0 \; , \end{array}$$

where $\dot{s}_0 \equiv \dfrac{ds_0}{dt}$ and $s_0(t)$ must be determined as part of the solution.

(2) Interior layer 1 (kink region).

Writing $u = \omega^{\frac{1}{3}} u^*$, $x = s(t;\omega) + \omega^{\frac{1}{3}} z$ with
$$u^* \sim u_0^*(z,t), \quad s \sim s_0(t) \quad \text{as } \omega \to 0$$
gives (cf. (4.7))

(4.13) $\quad -\dot{s}_0 u_0^* = \left(\dfrac{4 f_I f_V \, u_0^*}{(f_I - f_V) u_s} + \dfrac{h_I(f_I - f_V) u_s}{f_I \, u_0^{*2}} \right) \dfrac{\partial u_0^*}{\partial z}$,

and it is matching with this that leads to the condition (4.12). From (4.13) we obtain

(4.14) $\quad -\dot{s}_0 z = \dfrac{4 f_I f_V u_0^*}{(f_I - f_V) u_s} - \dfrac{h_I(f_I - f_V) u_s}{2 f_I u_0^{*2}}$

where the arbitrary function of t which arises on integrating (4.13) has been set to zero. This may be achieved by appropriate specification of the $O(\omega^{\frac{2}{3}})$ term in $s(t;\omega)$; completing the determination of s to this order requires the matching of further terms in the expansion for u, however.

It follows from (4.14) that
$$\text{as } z \to +\infty \quad u_0^* \sim \left(h_I(f_I - f_V) u_s \big/ 2 f_I \dot{s}_0 z \right)^{\frac{1}{2}}.$$

(3) Interior layer 2.

We write
$$u = \omega^{\frac{1}{2}} \ln^{\frac{1}{2}}(1/\omega) \bar{u}, \quad x = s(t;\omega) + \dfrac{\bar{z}}{\ln(1/\omega)};$$
these scalings are necessary to match into region (4). At leading order we have

(4.15) $\quad -\dot{s}_0 \bar{u}_0 - \alpha(t) = \dfrac{h_I(f_I - f_V) u_s}{f_I \bar{u}_0^2} \dfrac{\partial \bar{u}_0}{\partial \bar{z}}$,

where $\alpha(t)$ remains to be determined. Equation (4.15) implies that

(4.16) $\quad \text{as } \bar{z} \to +\infty \quad \bar{u} \sim h_I(f_I - f_V) u_s \big/ f_I \alpha \bar{z}$.

(4) Transition region.

We now have
$$u = \omega^{\frac{1}{2}} \ln^{-\frac{1}{2}}(1/\omega) u^\dagger, \quad x = s + O(1),$$

with
$$0 = \frac{\partial}{\partial x}\left(u_0^{\dagger\,-2}\frac{\partial u_0^{\dagger}}{\partial x}\right),$$
so that matching with (4.16) yields

(4.17) $\qquad u_0^{\dagger} = h_I(f_I - f_V)u_s \big/ f_I\alpha(x - s_0).$

(5) Tail region.

In this final region we write
$$V = \omega\widehat{V}, \quad I = \widehat{I}/\omega, \quad u = \omega\widehat{u}, \quad x = \widehat{x}/\omega^{\frac{1}{2}}$$
and obtain at leading order

(4.18) $\qquad \begin{cases} \dfrac{\partial \widehat{u}_0}{\partial t} = -\dfrac{\partial^2}{\partial \widehat{x}^2}(h_I \widehat{I}_0), \\[6pt] \dfrac{\partial}{\partial t}(\widehat{I}_0 \widehat{u}_0/\nu) = \dfrac{\partial^2}{\partial \widehat{x}^2}\left((h_I + f_I\widehat{u}_0)\widehat{I}_0\right), \\[6pt] \widehat{V}_0 = 1/\widehat{I}_0. \end{cases}$

The required solution to (4.18) satisfies

(4.19)
$$\text{as } x \to 0^+ \;\; \widehat{u}_0 \sim \left(h_I(f_I - f_V)u_s t/f_I\right)^{\frac{1}{2}} \Big/ \widehat{x}\,\ln^{\frac{1}{2}}(1/\widehat{x}), \quad \widehat{V}_0 \sim f_I\widehat{u}_0 \big/ (f_I - f_V)u_s,$$
$$I_0 \sim (f_I - f_V)u_s \big/ f_I\widehat{u}_0.$$

Matching (4.19) with (4.17) then yields
$$\alpha = \left(h_I(f_I - f_V)u_s \big/ 2f_I t\right)^{\frac{1}{2}}.$$

The discussion at the end of section 4.1 considered asymptotics on the diffusivity D_2 but not on the corresponding profile of u. If the latter were considered, the regions (1) – (4) would be essentially as in this section, while region (5) would simplify to give the diffusivity (4.9) (this arises if $h_I \gg 1$ in (4.18)). We note that (4.18) also corresponds to a limit in which flux-balance does not occur.

5. **Higher dimensions.** We restrict the discussion here to the case $r = O(1/g)$. The appropriate generalisations of (2.8) – (2.10) are then

(5.1) $\qquad \nabla^2\!\left((g_V r_V + f_V u)V\right) = \nabla^2\!\left((g_I r_I + f_I u)I\right) = -\left(\kappa_0 r + \kappa_1 u + \dfrac{\kappa_2}{r}u^2\right)(1 - IV),$

(5.2) $\qquad \dfrac{\partial u}{\partial t} = \nabla^2\!\left((f_V V + f_I I)u\right)$

and we may immediately note the following. Schaake [11] claims that in any number of dimensions there is a flux-balance condition, which for (5.1) would read

$$\nabla\Big((g_V r_V + f_V u)V\Big) = \nabla\Big((g_I r_I + f_I u)I\Big) \; ;$$

this is however, other than in one dimension, a far more restrictive condition than (5.1) and it cannot in general hold. This implies that in dimensions higher than one it is not possible to obtain algebraic expressions (such as (3.1) and (3.2) or (4.2) and (4.3)) relating the defect concentrations to the dopant concentrations in a simple local manner; the only non-local effects in (3.1) – (3.2) and (4.2) – (4.3) arise from the dependence on the surface concentration u_s. This would seem to indicate that attempts to extend the Fair–Tsai model to higher dimensions, such as [6] and [2], are largely inappropriate. Instead, coupled elliptic-parabolic systems such as (5.1) – (5.2) should be solved; such reduced systems do not have the extreme stiffness associated with the original systems.

We shall restrict further attention here to the form of the far-field behaviour associated with the systems (5.1) and (5.2). If κ_0, κ_1 and κ_2 are sufficiently small then (5.1) reduces to

$$(5.3) \qquad \nabla^2\Big((g_V r_V + f_V u)V\Big) = \nabla^2\Big((g_I r_I + f_I u)I\Big) = 0 \; ,$$

but when the recombination-generation terms dominate in the far-field we should consider

$$(5.4) \qquad \nabla^2\Big((g_V r_V + f_V u)V\Big) = \nabla^2\Big((g_I r_I + f_I u)I\Big), \qquad IV = 1 \; .$$

We shall attempt to be as general as possible with regard to boundary and initial conditions, subject to the following constraints. We consider a two-dimensional problem on the domain

$$-\infty < y < +\infty \; , \qquad 0 \leq x < +\infty \; ,$$

$x = 0$ being the silicon surface. We assume that

$$u \to 0 \quad \text{as } x \to +\infty \quad \text{or} \quad \text{as } y \to -\infty$$

and that the behaviour is one-dimensional in the limit $y \to +\infty$, and can therefore be described by the analysis given earlier. Such conditions are applicable to the important problem of diffusion under a mask edge.

It then follows that as $x \to +\infty$ with $y/x \to +\infty$ we have

$$(5.5) \qquad V \to V_\infty(t) \; , \qquad I \to I_\infty(t) \; ,$$

where for (5.3) V_∞ and I_∞ are given by (3.6), while for (5.4) we have

$$(5.6)$$
$$V_\infty = 1/I_\infty, \qquad I_\infty = \Big\{\big((g_I r_I - g_V r_V + (f_I - f_V)u_s)^2 + 4g_I r_I g_V r_V\big)^{\frac{1}{2}}$$
$$+ (g_I r_I - g_V r_V + (f_I - f_V)u_s)\Big\}\Big/2g_I r_I \; .$$

We note that in either case we have

(5.7) $$g_V r_V V_\infty - g_I r_I I_\infty = g_V r_V - g_I r_I + (f_V - f_I) u_s ;$$

this follows from (2.8). In addition, for any x we have

(5.8) $$\text{as } y \to -\infty \quad V \to 1, \quad I \to 1 .$$

Writing $x = r \cos\theta$, $y = r \sin\theta$, the far-field defect behaviour as $x \to +\infty$ with $y/x = O(1)$ may now be obtained. Since $u \to 0$ in this limit it follows from (5.1), (5.7) and (5.8) that

(5.9) $$g_V r_V V - g_I r_I I \sim g_V r_V - g_I r_I + \frac{1}{2}(f_V - f_I)\left(1 + \frac{2\theta}{\pi}\right) u_s ,$$

and for (5.3) we obtain

(5.10) $$V \sim 1 + \frac{1}{2} s_V \left(1 + \frac{2\theta}{\pi}\right) u_s , \quad I \sim 1 + \frac{1}{2} s_I \left(1 + \frac{2\theta}{\pi}\right) u_s ,$$

while for (5.4) we have $V = 1/I$ with

$$I \sim I^*(\theta, t)$$

where

(5.11)
$$I^* = \left\{ \left((g_I r_I - g_V r_V + \frac{1}{2}(f_I - f_V)\left(1 + \frac{2\theta}{\pi}\right) u_s)^2 + 4 g_I r_I g_V r_V \right)^{\frac{1}{2}} \right.$$
$$\left. + \left(g_I r_I - g_V r_V + \frac{1}{2}(f_I - f_V)\left(1 + \frac{2\theta}{\pi}\right) u_s \right) \right\} / 2 g_I r_I .$$

To determine the resulting far-field behaviour of the dopant concentration u we introduce an artificial small parameter δ by writing

$$R = \delta r .$$

We now define

$$\phi(\theta, t) = 1 + \frac{1}{2}(f_V s_V + f_I s_I)\left(1 + \frac{2\theta}{\pi}\right) u_s$$

for (5.3) and

$$\phi(\theta, t) = f_I I^* + f_V / I^*$$

for (5.4) (we note that ϕ depends on θ and t only through the combination $\left(1 + \frac{2\theta}{\pi}\right) u_s(t)$). Equation (5.2) then implies that in either case

(5.12) $$\frac{\partial u}{\partial t} \sim \frac{1}{\delta^2} \left(\frac{1}{R} \frac{\partial}{\partial R} \left(\frac{\partial}{\partial R}(\phi u) \right) + \frac{1}{R^2} \frac{\partial^2}{\partial \theta^2}(\phi u) \right) .$$

and we apply the W.K.B.J. method by assuming that

(5.13) $$u \sim \Gamma(\delta)\Big(a_0(R,\theta,t) + \cdots\Big)e^{-F(R,\theta,t)/\delta^2} \quad \text{as } \delta \to 0,$$

for some function $\Gamma(\delta)$.

The far-field behaviour is largely determined by F, which satisfies the first order partial differential equation

$$\frac{\partial F}{\partial t} = -\phi(\theta,t)\left(\left(\frac{\partial F}{\partial R}\right) + \frac{1}{R^2}\left(\frac{\partial F}{\partial \theta}\right)^2\right).$$

Since δ is an artificial small parameter, the require solution takes the form

$$F = R^2 G(\theta,t)$$

with

(5.14) $$\frac{\partial G}{\partial t} = -\phi(\theta,t)\left(\left(\frac{\partial G}{\partial \theta}\right)^2 + 4G^2\right).$$

By solving (5.14) by the method of characteristics we may, for example, determine the ratio of lateral to vertical diffusion lengths without having to consider the full system (5.1) – (5.2). The appropriate boundary condition on (5.14) is obtained by matching into the one-dimensional region corresponding to $y \to +\infty$; this implies that

(5.15) $$\text{as } \theta \to \frac{\pi}{2}^{-} \quad G \sim \left(\theta - \frac{\pi}{2}\right)^2 \Big/ 4 \int_0^t \phi\left(\frac{\pi}{2}, u_s(t')\right) dt'.$$

If u_s is a constant, so that ϕ does not depend on t, then the problem may be further reduced to a single ordinary differential equation because G takes the form

$$G(\theta,t) = H(\theta)/t$$

with

$$\left(\frac{dH}{d\theta}\right)^2 + 4H^2 = H/\phi(\theta),$$

$$\text{as } \theta \to \frac{\pi}{2}^{-} \quad H \sim \left(\theta - \frac{\pi}{2}\right)^2 \Big/ 4\phi\left(\frac{\pi}{2}\right).$$

6. Discussion. The paper has outlined the results of applying singular perturbation methods to a simple model for dopant-defect pair diffusion. Much of the analysis carries over to more realistic models which allow for the electric charges carried by the various species.

Some of the reduced problems given here are not new, but have been obtained before by physical reasoning; see [8] and [9] in particular. We are, however, able to state precise conditions on the governing parameters in order for such reductions to be valid. In particular, the reduced problems which follow from flux-balance, namely (3.3) – (3.4) and (4.2) – (4.5) (see also [8] and [9]), require that, for example, $g_I r_I = O(1)$ or in dimensional terms that

$$c_d^* = O\left(c_I^* D_I \big/ (D_{dV}\omega_V + D_{dI}\omega_I)\right).$$

The higher concentration problems (see sections 3.2 and 4.2), which are new, are appropriate when, for example, $g_I r_I = O(\omega_I)$ so that

$$c_d^* = O\left(c_I^* D_I \big/ \omega_I(D_{dV}\omega_V + D_{dI}\omega_I)\right).$$

In these cases a fuller balance of terms occurs (see (3.8) and (4.18)), but the surface region can take a particular simple form (see (3.9)).

The reduced problems in one dimension often take the form of nonlinear diffusion equations which contain a non-local dependence on the surface concentration (this surface dependence arises from the formation of pairs at the surface which diffuse in and then dissociate); see (3.3) and (4.4). Some general considerations for such equations are given in [5]. A crucial step in the derivation of such models is the balancing of fluxes of pairs and defects. As we have shown, such flux-balance conditions do not in general hold for high concentrations or in more than one dimension. For this reason extensions of simplified one-dimensional models into higher dimensions should be treated with caution. In higher dimensions the dopant diffusion equations arising from the asymptotic analysis exhibit not only non-local dependence but also direction dependence. This is exemplified by (5.12).

The results given at the end of section 4.1 illustrate how effective diffusivities which may explain the form of diffused phosphorus profiles may be derived if both interstitial and vacancy effects are considered (cf. [9] and [10]). For the surface region diffusivity (4.6) to be sufficiently large we require that neither f_I nor f_V be too small, so that both mechanisms must be operative. It is nevertheless possible for f_I to be significantly larger than f_V (or vice versa) so that under low concentration conditions the mechanism of diffusion may be dominated by a particular point defect.

REFERENCES

[1] P.M. FAHEY, P.B. GRIFFIN AND J.D. PLUMMER, *Point defects and dopant diffusion in silicon*, Rev. Mod. Phys., **61** (1989), pp. 289–384.

[2] R.B. FAIR, C.L. GARDNER, M.J. JOHNSON, S.W. KENKEL, D.J. ROSE, J.E. ROSE AND R. SUBRAHMANYAN, *Two-dimensional process simulation using verified phenomenological models*, IEEE Trans. Comp.-Aided Des., **10** (1991), pp. 643–650.

[3] R.B. FAIR AND J.C.C. TSAI, *A quantitative model for the diffusion of phosphorus in silicon and the emitter dip effect*, J. Electrochem. Soc., **124** (1977), pp. 1107–1118.

[4] J.R. KING, *Asymptotic analysis of an impurity-defect pair diffusion model*, Q.J. Mech. Appl. Math., **44** (1991), pp. 369–412.

[5] J.R. KING, *Surface-concentration dependent nonlinear diffusion*, Euro. J. Appl. Math (to appear).

[6] F. LAU AND U. GÖSELE, *Two-dimensional phosphorus diffusion for soft drains in silicon MOS transistors*, Appl. Phys. A, **40** (1986), pp. 101–107.

[7] J.W. MOORE AND R.G. PEARSON, *Kinematics and mechanism*, John Wiley, New York, (1981).

[8] F.F. MOREHEAD AND R.F. LEVER, *Enhanced "tail" diffusion of phosphorus and boron in silicon: self-interstitial phenomena*, Appl. Phys. Lett., **48** (1986), pp. 151–153.

[9] F.F. MOREHEAD AND R.F. LEVER, *The steady-state model for coupled defect-impurity diffusion in silicon*, J. Appl. Phys., **66** (1989), pp. 5349–5352.

[10] W.B. RICHARDSON AND B.J. MULVANEY, *Plateau and kink in P profiles diffused into Si: a result of strong bimolecular recombination?*, Appl. Phys. Lett., **53** (1988), pp. 1917–1919.

[11] H.F. SCHAAKE, *The diffusion of phosphorus in silicon from high surface concentrations*, J. Appl. Phys., **55** (1984), pp. 1208–1211.

A REACTION-DIFFUSION SYSTEM MODELING PHOSPHORUS DIFFUSION

WALTER B. RICHARDSON, JR.[*]

Abstract. At very high concentrations phosphorus diffusion in silicon exhibits marked nonlinearities. The hierarchy of physical models that attempt to explain this anomalous diffusion are reviewed. An eight-species kinetic model is derived that yields a quasilinear, partly-dissipative system of reaction-diffusion partial differential equations. The numerical method of lines is used to solve the system for a simplified five-species model in three dimensions. The linear system in the Newton iteration is solved using several matrix-free methods. In all cases the dimension of the Krylov subspace must be quite large to insure convergence. This suggests that preconditioning will be more important for efficiency than choice of an accelerator.

Key words. nonlinear diffusion, reaction-diffusion, semiconductor doping, method of lines.

AMS(MOS) subject classifications. 65J15,60J60,35K57

1. Introduction. The trend toward shrinking device dimensions in Very Large Scale Integration has produced an increased need for accurate simulation tools for Technology Computer Aided Design. This has lead to a hierarchy of increasingly sophisticated models for device simulation including drift-diffusion, hydrodynamic, and full Boltzmann. These tools rely upon process simulation for input data, including accurate impurity concentration profiles and material boundaries. Physical processes involved in fabricating an integrated circuit include lithography, etching, deposition, diffusion, ion implantation, epitaxy, and oxidation. Process modeling is less mature than device modeling, in part because the physics is less well understood. It provides a wealth of open problems for mathematicians interested in nonlinear diffusion, moving boundary problems, and large reaction-diffusion-convection systems.

This paper treats only the problem of phosphorus diffusion in silicon. The model used currently and several of its refinements are reviewed. Experimental evidence shows that phosphorus diffuses by a dual interstitialcy-vacancy mechanism. An eight-species kinetic model is derived. Using experimentally determined estimates for the diffusivities and simple kinetic approximations for the rate constants, it is shown that numerical simulations of a high concentration predeposition do exhibit the anomalous plateau, kink, and tail, characteristic of phosphorus diffusion. This model has been implemented in 3-D using the numerical method of lines with the system integrator LSODP. Results on several test problems indicate that for its linear solver, SIOM, as well as several of the accelerators in the NSPCG package, the size of the Krylov subspace must be taken sufficiently large to insure convergence in the inner Newton iteration. Finally, analytic results for the partly-dissipative system are reviewed and directions for future work given.

2. Nonlinear Diffusion. High temperature diffusion is used to introduce

---

[*] Department of Mathematics, University of Texas at San Antonio, San Antonio, Texas 78249. This work was supported in part by NSF Grant DMS - 9024712

dopants such as phosphorus, arsenic, and boron into a silicon wafer in order to form n-p junctions, as well as to anneal damage to the lattice caused by ion implantation. When the impurity concentration C is less than the intrinsic electron concentration n_i, approximately 10^{18}cm^{-3} at 1000°C, the heat equation represents dopant diffusion well. For predeposition it is solved with a Dirichlet boundary condition at the top surface of the wafer

(1) $\quad C_t = D \Delta C \qquad C(0,t) = C^* \text{ and } \dfrac{\partial C}{\partial \xi}(\infty, t) = 0 \qquad C(x,0) = 0$

As the ambient concentration C^* increases, measured diffusivities show a marked concentration dependence. To explain this, Schwettmann, Yoshida, and others postulated that diffusion was more than a simple random-walk nearest neighbor interchange mechanism. It is thermodynamically more favorable for an impurity atom to bond with a vacancy, a lattice site not occupied by a silicon atom, and the pair move through the lattice. Vacancies exist in several charge states, $V^0, V^+, V^-, V^=$, with a density that is temperature and Fermi-level dependent. Diffusion depends on the concentration of vacancies and the concentration of, say, acceptor vacancies $[V^-]$ is proportional to the electron concentration n. Assuming charge neutrality, n depends on C as in $n = \frac{1}{2}\left(C + \sqrt{C^2 + 4n_i^2}\right)$; which gives a nonlinear diffusion equation

(2) $\qquad \dfrac{\partial C}{\partial t} = \nabla \cdot \{\, D(C)\, [\, \nabla C \;+\; C\, \nabla(\ln n)\,]\,\}$

Here $D(C)$ is a compound diffusivity

(3) $\qquad D(C) \;=\; D_0^i \;+\; D_+^i \left(\dfrac{n_i}{n}\right) \;+\; D_-^i \left(\dfrac{n}{n_i}\right) \;+\; D_=^i \left(\dfrac{n}{n_i}\right)^2$

and D_0^i, D_+^i, D_-^i, and $D_=^i$ are the intrinsic diffusivities due to neutral, donor, acceptor, and double acceptor vacancies, respectively. If more than one impurity is present, there would be additional equations for the other impurities, and in the drift term the concentration C is replaced by the net electrically active concentration N. Using (2) gives better results for extrinsic (high concentration) diffusion than (1), but still fails to explain the pronounced nonlinearities that occur for a high-concentration phosphorus source.

In addition to vacancies, interstitials – silicon atoms not residing on a lattice site – are known to exist in numbers roughly equal to vacancies. Experiments such as oxidation enhanced diffusion suggest that these species also aid in the diffusion process. In 1974, Hu proposed that "P diffuses in Si via a dual mechanism, i.e., a mixture of vacancy and interstitialcy mechanisms," by means of the reactions

$$V \;+\; P_i \rightleftharpoons P_s \;+\; S$$

$$S \;+\; P_i \rightleftharpoons P_s \;+\; I$$

where S denotes a Si lattice atom, P_s (P_i) substitutional (interstitial) phosphorus, and V (I), a vacancy (interstitial), respectively. Over the next fifteen years a number of models were developed that refined this idea, including those of Mathiot-Pfister, Morehead-Lever, Law-Dutton, and Mulvaney-Richardson. They solved for concentrations of as many as three of the chemical species, using various assumptions about equilibrium to simplify the model.

3. Diffusion-Limited Kinetics. A completely nonequilibrium model [1],[2] was formulated in terms of the five reactions

$$V^0 + I^0 \rightleftharpoons <0> \quad (R1)$$
$$P^+ + V^- \rightleftharpoons P^+V^- \quad (R2) \qquad P^+ + I^- \rightleftharpoons P^+I^- \quad (R3)$$
$$V^0 + e^- \rightleftharpoons V^- \quad (R4) \qquad I^0 + e^- \rightleftharpoons I^- \quad (R5)$$

This results in the quasilinear reaction diffusion system,

$$\frac{\partial [e^-]}{\partial t} = -k_4^f [e^-][V^0] + k_4^r [V^-] - k_5^f [e^-][I^0] + k_5^r [I^-]$$

$$+ k_{neu} \{ [P^+] + \frac{n_i^2}{[e^-]} - [e^-] - [V^-] - [I^-] \}$$

$$\frac{\partial [V^0]}{\partial t} = \nabla \cdot \{ D_{V^0} \nabla [V^0] \} - k_4^f [e^-][V^0] + k_4^r [V^-]$$

$$- k_{bi} ([I^0][V^0] - [I^0]^{eq}[V^0]^{eq})$$

(4) $$\frac{\partial [V^-]}{\partial t} = \nabla \cdot \{ D_{V^-} \left(\nabla [V^-] - [V^-] \nabla \ln(n) \right) \}$$

$$+ k_4^f [e^-][V^0] - k_4^r [V^-] - k_2^f [P^+][V^-] + k_2^r [P^+V^-]$$

$$\frac{\partial [P^+V^-]}{\partial t} = \nabla \cdot \{ D_{P^+V^-} \nabla [P^+V^-] \} + k_2^f [P^+][V^-] - k_2^r [P^+V^-]$$

$$\frac{\partial [P^+]}{\partial t} = -k_2^f [P^+][V^-] + k_2^r [P^+V^-] - k_3^f [P^+][I^-] + k_3^r [P^+I^-]$$

with three analogous equations for interstitial species. Reaction R1 represents bimolecular generation-recombination with rate constant k_{bi}. $<0>$ denotes the result of an interstitial silicon atom occupying a lattice site and annihilating the vacancy. For reactions R2-R5, the forward and reverse rate constants are denoted by k_n^f and k_n^r respectively. In the equation for electron concentration (denoted both by $[e^-]$ and n), the last term dynamically enforces charge neutrality.

The single species model with a nonlinear diffusivity has been exchanged for a large system each equation of which is quasilinear with constant diffusivity. The kinetic model attempts to more accurately embody the physics and gives data about species whose concentrations cannot be measured directly. This is at the expense of more equations and a stiffer system. Boundary conditions on the upper surface are

(5) $$\frac{\partial [e^-]}{\partial \vec{\xi}} = \frac{\partial [P^+V^-]}{\partial \vec{\xi}} = 0$$

(6) $\qquad [V^0] = [V^0]^{eq} \qquad [V^-] = (\frac{n}{n_i}) \cdot [V^-]^{eq} \qquad [P^+] = C^*$

with, of course, analogous conditions for the interstitial species. In the Dirichlet condition for P^+, C^* represents the concentration of phosphorus in the ambient gas.

Reflecting Neumann conditions are enforced for all species deep in the bulk. Initial conditions are zero for all species except the electrons and defects, which are set to their equilibrium levels.

For vacancies it is possible to estimate the diffusivities from first principles using thermodynamics. For interstitials and pairs the data is nonexistent or of doubtful accuracy, and experimental data of Mathiot-Pfister was used to give the following values

Table I Diffusivities at 900°C $(cm^2 s^{-1})$.

D_V	D_I	D_{P+V-}	D_{P+I-}
4.0×10^{-8}	5.8×10^{-8}	1.4×10^{-12}	6.2×10^{-12}

For the reaction $A + B \rightleftharpoons C$, a simple kinetic argument due to Debye gives an estimate for the forward rate constant for a diffusion-limited reaction of

$$(7) \qquad k^f = 4 \pi R (D_A + D_B)$$

where R is the encounter distance for A-B interaction. The rate constant for the reverse reaction is approximated by

$$(8) \qquad k^r = k^f \frac{1}{4} n_h \exp(- E_b / kT)$$

where n_h is the concentration of lattice sites and E_b is the binding energy of the $A - B$ pair. For reactions R4-R5 involving electrons this is precisely analogous to Schockley-Read-Hall theory of recombination-generation in which

$$(9) \qquad k^r = v_{th} \sigma_n n_i \exp(E_{V-} - E_i)$$

where $v_{th} \approx 10^7$ cm/sec is the thermal velocity of an electron, $\sigma_n \approx 10^{-15} cm^2$ is the capture cross-section, and E_{V-} is the energy level of the acceptor vacancy.

Table II Reaction Constants at 900°C.

Reaction	R2	R3	R4	R5
Forward ($cm^3 s^{-1}$)	3.0×10^{-14}	4.4×10^{-14}	1.0×10^{-8}	1.0×10^{-8}
Reverse (s^{-1})	1.4×10^2	8.8×10^1	5.6×10^{10}	1.7×10^{11}

In order to obtain a marked plateau, the bimolecular generation-recombination rate, k_{bi}, was initially taken to be $1.7 \times 10^{-7} cm^3 s^{-1}$, which is several orders of magnitude greater than the one determined from (7). Using an encounter distance for vacancy-interstitial interaction of roughly 10^{-7} cm, the values of D_V and D_I from Table I give an estimate of $7 \times 10^{-14} cm^3 s^{-1}$ for k_{bi}. It has been shown, however, that recombination is strongly dependent on impurity concentration and this dependence will result in a much larger effective rate. Note that there are four important alternate paths for defect recombination

$$P^+ V^- + I^0 \rightarrow P^+ + e^- \qquad (R6) \qquad P^+ I^- + V^0 \rightarrow P^+ + e^- \qquad (R7)$$
$$P^+ V^- + I^- \rightarrow P^+ + 2e^- \qquad (R8) \qquad P^+ I^- + V^- \rightarrow P^+ + 2e^- \qquad (R9)$$

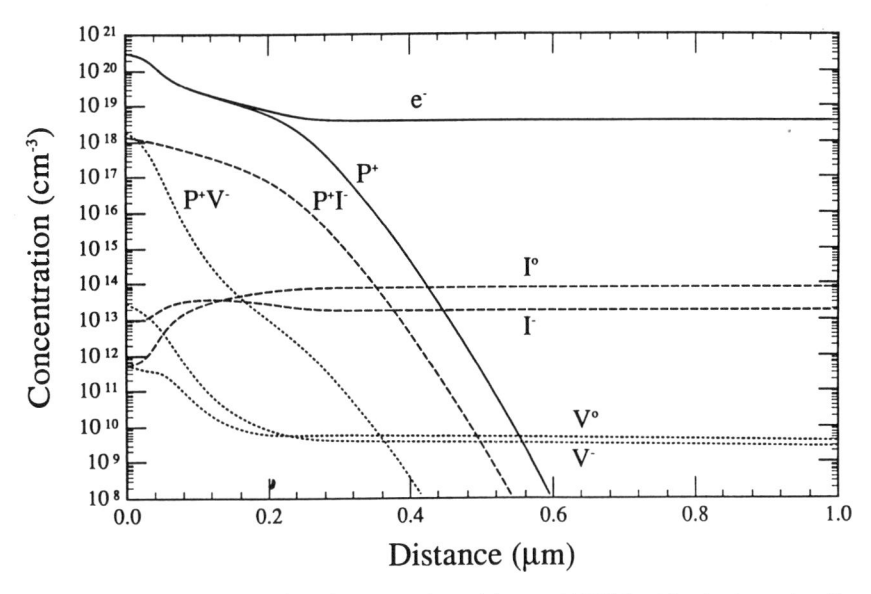

Figure 1: Simulation of a phosphorus predeposition at 900°C for 10 minutes using the eight-species nonequilibrium model in one dimension. This includes the additional terms from reactions R6-R9. Crosses are experimental data of Yoshida *et al., J. Appl. Phys.* p. 1498(1974).

Inclusion of R6 in the model would result in an additional term of the form $-k_6^f [P^+V^-][I^0]$ in the equation for I^0. At equilibrium

$$[P^+V^-] = \frac{k_2^f}{k_2^r}[P^+][V^-] = 2 \times 10^{-16}[P^+]\left(\frac{n}{n_i}\right)[V^0]^{eq}$$

Given a surface phosphorus concentration of 3×10^{20}, $[P^+V^-] \approx 10^6 [V^0]$ for the region close to the surface and

$$k_6^f [P^+V^-][I^0] \approx 10^6 k_6^f [V^0][I^0]$$

For the reaction of P^+V^- and I^0, orientation effects would be important as compared with the direct recombination mechanism. Still, supposing that $k_6^f \approx k_{bi}$, k_{bi}^{eff} would be 10^6 times the value predicted from (7).

Numerical modeling using kinetic estimates for all rate constants, ($k_{bi} = 10^{-14}$ cm^3 s^{-1}) and including reactions R6-R9 gives Figure 1 [3]. Note the pronounced plateau, kink, and tail in the profile. Comparison between this and the five-species model presented below suggests that the effect of charge is secondary. The primary reason for the observed anomalies is the interaction between the two types of defects and pairs via reactions R1, R6-R9.

4. Numerical Solutions in Three Dimensions. The above diffusion models were among several implemented in the one-dimensional process simulator PEPPER

[4] developed at Microelectronics and Computer Technology Corporation during 1985-1986. Work to extend these models to three dimensions is ongoing [5]. Results are obtained for both the standard model (2) and a five-species kinetic model that results from simplifying (4):

$$\frac{\partial V}{\partial t} = D_V \nabla^2 V - k_E^f P \cdot V + k_E^r E - k_{bi} (V \cdot I - V^{eq} \cdot I^{eq})$$

$$\frac{\partial I}{\partial t} = D_I \nabla^2 I - k_F^f P \cdot I + k_F^r F - k_{bi} (V \cdot I - V^{eq} \cdot I^{eq})$$

(10) $$\frac{\partial E}{\partial t} = D_E \nabla^2 E + k_E^f P \cdot V - k_E^r E$$

$$\frac{\partial F}{\partial t} = D_F \nabla^2 F + k_F^f P \cdot I - k_F^r F$$

$$\frac{\partial P}{\partial t} = - k_E^f P \cdot V + k_E^r E - k_F^f P \cdot I + k_F^r F$$

Here E represents the phosphorus-vacancy pair or E-center and F denotes P^+I^-. The partial differential equations are discretized using the numerical method of lines. This continuous-in-time, discrete-in-space method converts (10) into a large coupled nonlinear system of ordinary differential equations, which is then integrated using the system integrator package LSODP (Livermore Solver for Ordinary Differential equations - Projection), developed by Brown and Hindmarsch [6]. NMOL has proven particularly effective on large reaction-diffusion problems; less effective when strong convective terms are present. Members of the LSODE family are variable-order, variable-step and use a predictor-corrector scheme based on Backward Differentiation Formulae to control error. This implicit method requires the solution of a nonlinear algebraic system at each time step which is performed via a Quasi (an approximate Jacobian that is only updated as needed), Inexact (the linear system is only approximately solved using the Scaled Incomplete Orthogonalization Method) Newton iteration. The ODE system is extremely stiff because: 1) there are over four orders of magnitude difference in the diffusivities of the various species as seen from Table I, 2) a very fine grid is required to resolve the sharp gradients in defect concentrations near the surface of the wafer and it is the grid spacing which determines the eigenvalues of the discrete Laplacian, and 3) the strong reaction kinetics apparent from Table II. For convenience the simulation domain chosen is a rectangular cube with reflecting boundary conditions on all but the top side where the conditions (5)–(6) are enforced. Finite differences with a seven point star discretize the Laplacian and pointwise estimates are used for the reaction terms. The absolute and relative tolerances are set at 10^{10} and 10% respectively and, unless otherwise noted, the remaining LSODP parameters take their default values. A suite of 3 test problems is used with Problem A possessing symmetry in two directions so that it is pseudo 1-D. Figure 2 shows the geometries for Problems B and C.

The SIOM algorithm [7] has very modest storage requirements and is "matrix free" requiring only the matrix-vector product Av for a given v. To solve $Ax = b$ it takes a shifted Krylov subspace K_m of $I\!R^n$ and seeks an approximate solution $x^{(m)}$ which belongs to K_m and such that the residual $r^{(m)}$ at $x^{(m)}$ is orthogonal to K_m. An Arnoldi method recursively builds an orthonormal basis $\{v_1, ..., v_m\}$ for K_m; incomplete refers to the fact that v_k is only orthogonalized against the previous $\{v_{k-1}, ..., v_{k-j}\}$. Saad

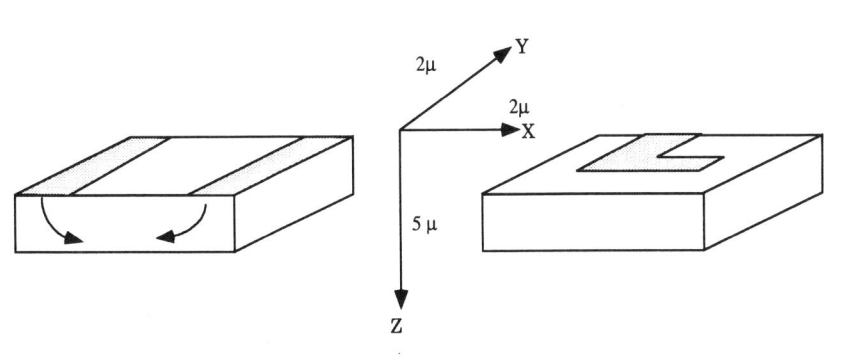

Figure 2: Geometries for the three dimensional test problems. B possesses symmetry along one axis and can be compared with output from a 2-D simulator. C represents a problem that would be difficult to model using a sequence of 2-D simulations.

has shown that when **A** is symmetric, SIOM reduces to conjugate gradient and in general is equivalent to ORTHORES and ORTHOMIN.

Good results are achieved with the standard model with grids of up to 27,000 nodes. The kinetic model results in a much stiffer system, and to get convergence on even the simple Problem A, the maximum dimension of the Krylov subspace must be taken much larger than its default value of $m = 5$. Figure 3 shows the effect on vacancy and interstitial profiles of changing the maximum Krylov dimension. The integrator halted due to nonconvergence with m less than eight. For $m = 10$, a solution is obtained, but it is unphysical and radically different from a 1-D PEPPER simulation where a direct solver is used. Refining the grid does not improve the situation: in Figure 4 a grid of 80 points is used in the z direction and m must be 30 before relatively flat profiles are obtained. If the Krylov subspace is large enough for convergence, it may still be insufficient to produce a physically meaningful solution. It is important to solve the linear system accurately at each timestep and this cannot be done simply by reducing the tolerance in the outer loop, one must enlarge the subspace or cut back on the stepsize severely. For the SIOM algorithm in LSODP the quantity **A**v's/Calls is the average Krylov dimension and represents a measure of the efficiency of the method. Figure 5 shows K_{avg} for both models on a grid of 1000 points. As time increases the asymmetric Jacobian makes a larger contribution to the linear system, effectively requiring a larger Krylov subspace to obtain an accurate solution.

To compare SIOM with other accelerators, the calling sequence in LSODP is modified so that routines from the NonSymmetric Preconditioned Conjugate Gradient

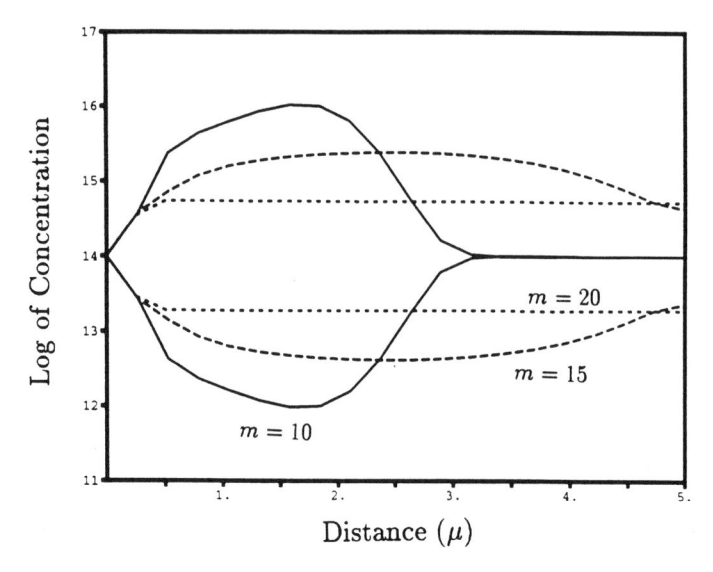

Figure 3: The effect on vacancy and interstitial profiles of increasing the Krylov subspace dimension. This is Problem A for five species with 20 gridpoints in the z direction. The lower curve of each pair is the vacancy, the upper the interstitial.

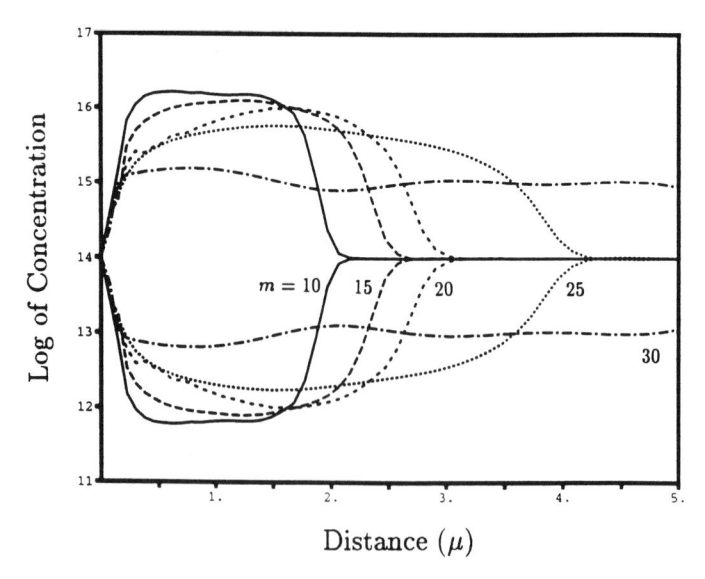

Figure 4: Same problem as in Figure 3, but with 80 gridpoints in the z direction. $Dim(K_m)$ must be as large as 30 to achieve the expected flat defect profiles.

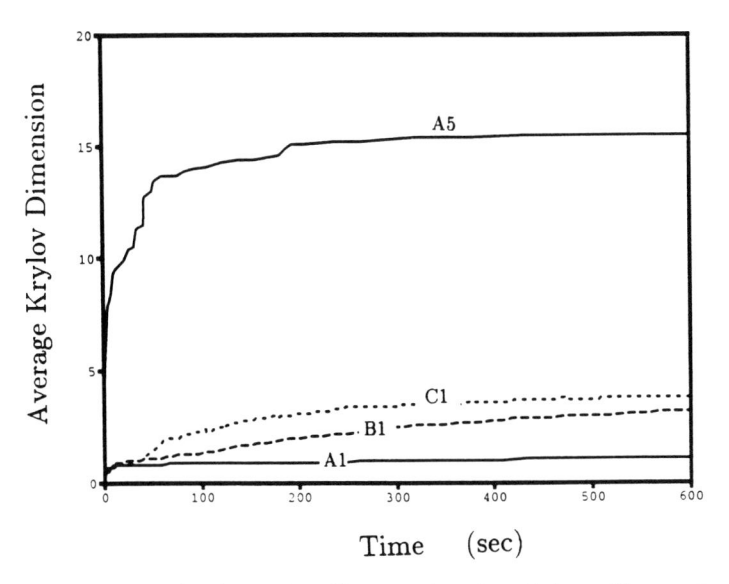

Figure 5: Average Krylov dimension K_{avg} as a function of time. The number K_{avg} is a measure of the relative effectiveness of the SIOM algorithm in solving the linear system.

package [8] can be called. It uses various accelerator techniques such as Chebyshev and generalized conjugate gradient. Four accelerators were chosen for comparison, ORTHOMIN, GMRES, BCGS, and Minimum Error. Table III gives the runtimes in seconds on a Sparcstation for the standard model (2) on a grid of one thousand points, where F(u)'s is the number of evaluations of the RHS in (10), $\mathbf{A}\mathbf{v}$'s is the number of times the matrix-vector product is formed, and Calls is the number of calls made to the iterative solver. All four algorithms performed well, with roughly equivalent runtimes, although GMRES and ORTHOMIN were slightly faster.

TABLE III.

Prob	Method	Time	F(u)'s	$\mathbf{A}\mathbf{v}$'s	Calls
A	BCGS	132.7	314	244	69
	GMRES	109.3	253	183	69
	ME	135.2	314	244	69
	ORTHOMIN	107.7	253	183	69
B	BCGS	1452.0	3489	2784	704
	GMRES	1087.4	2589	1935	653
	ME	1448.3	3489	2784	704
	ORTHOMIN	1167.3	2793	2088	704
C	BCGS	1898.4	4544	3628	915
	GMRES	1555.7	3661	2739	921
	ME	1899.9	4544	3628	915
	ORTHOMIN	1537.6	3637	2721	915

Of more importance than the choice of a particular accelerator will be selection of an effective preconditioner.

5. Conclusions and Future Work. Process modeling, and in particular modelling diffusion of impurities, is less mature than device modeling and represents an area rich in open problems for applied mathematicians. As the trend towards Ultra Large Scale Integration continues, there will be an even greater need for more accurate physical models and numerical techniques to model diffusion, oxidation, lithography, and vapor deposition. Detailed theoretical analyses of the reaction-diffusion-convection equations that arise in process simulation are just beginning. King [9] performs an asymptotic analysis for the standard model and the phenomenological Fair-Tsai model for phosphorus. Yeager [10] examines a "reactive-definite" system involving several species but only one reaction,

$$(11) \qquad u_t^i = \Delta u^i \pm f(u_1, ..., u_n)$$

including the question of convergence of the discrete Newton method. Hollis and Morgan [12] prove that the five-species system (10) satisfies their theory of partly-dissipative reaction-diffusion systems

$$(12) \qquad U_t = D\Delta U + F(U)$$

where $D = Diag(d_1, ..., d_m)$ with $d_i \geq 0$, because the function F corresponding to (10) satisfies conditions of 1) quasipositivity 2) the intermediate sum condition of Morgan and 3) polynomial growth. They observe that the boundary condition in (6) for substitutional phosphorus should be replaced by Dirichlet conditions for the pairs, since P^+ has zero diffusivity. From a numerical standpoint either set of conditions is enforced via a penalty method and chemical equilibrium is very quickly achieved at the wafer surface, so that similar results are obtained.

We have presented a hierarchy of models for phosphorus diffusion in Si, culminating with an eight-species reaction-diffusion system. It has been shown that this model can be implemented in three dimensions using current technology on grids of 20^3 points. As the mesh is further refined, runtimes become prohibitive for all the iterative solvers and for both models. This is explained by noting that when the Laplacian is discretized, the resulting matrix **A** has a spectrum that depends on the mesh spacing. **A** becomes increasingly ill-conditioned as the meshsize goes to zero. This carrys over to the operators $\nabla \cdot (D(C)\nabla C)$ and $D\Delta U$, and gives rise to stiffer ODE systems. Our results suggest that preconditioning will be very important in solving these problems on realistic meshes. Future work will include preconditioning with the strategies of Bramble and also Neuberger. A detailed theory of convergence for the discretized system has yet to be worked out. This analysis should be made for very irregular grids yet with realistic boundary conditions such as injection of interstitials during oxidation. Precise *a posteriori* error estimates would allow for accurate grid refinement as part of the solution process.

REFERENCES

[1] W. B. Richardson and B. J. Mulvaney, *Plateau and kink in P profiles diffused into Si - A Result of Strong Bimolecular Recombination?*, Appl. Phys. Lett. 53(1988), pp. 1917-1919.

[2] W. B. Richardson and B. J. Mulvaney, *Nonequilibrium behavior of charged point defects during phosphorus diffusion in silicon*, J. Appl. Phys., 65(1989), pp. 2243-2247.

[3] B. J. Mulvaney and W. B. Richardson, *The effect of concentration-dependent defect recombination reactions on phosphorus diffusion in silicon*, J. Appl. Phys. 67(1990), pp. 3197-3199.

[4] B. J. Mulvaney, W.B. Richardson, and T. Crandle, *PEPPER - A Process Simulator for VLSI*, IEEE Trans. on Computer-Aided Design, 8(1989), pp. 336-349.

[5] W. B. Richardson, G. F. Carey, and B. J. Mulvaney, *Modeling Phosphorus Diffusion in Three Dimensions*, IEEE Trans. on Computer-Aided Design, 11(1992), pp. 487-496.

[6] P. N. Brown and A. C. Hindmarsch, SIAM J. Numer. Anal., 24(1987), pp. 610.

[7] Y. Saad, *Krylov Subspace Methods for Solving Large Unsymmetric Linear Systems*, Math. of Comp., vol. 37, 105(1981).

[8] T. C. Oppe, W. D. Joubert, and D. R. Kincaid, *Center for Numerical Analysis Report CNA-216*, University of Texas.

[9] J. R. King, *On the diffusion of point defects in silicon*, SIAM J. Appl. Math. 49(14), 1989, pp. 1018-1101.

[10] H. R. Yeager and R. W. Dutton, *An Approach to Solving Multiparticle Diffusion Exhibiting Nonlinear Stiff Coupling*, IEEE Trans. Electron Devices, vol. ED-32, pp. 1964-1976, 1985.

[11] S. L. Hollis, J. J. Morgan, and W. B. Richardson, *Partly Dissipative Reaction-Diffusion Systems and a Model of Phosphorus Diffusion in Silicon*, submitted.

ATOMIC DIFFUSION IN GaAs WITH CONTROLLED DEVIATION FROM STOICHIOMETRY

KEN SUTO* AND JUN-ICHI NISHIZAWA**

Abstract. Although several models for atomic diffusion in $GaAs$ have been presented, they have not given strong attention to the effect of the arsenic vapor pressure, i.e., the deviation from stoichiometry, or some of them are thought to be unrealistic.

On the other hand, we have shown that the crystal growth from solution or melt under applied vapor pressure i.e., temperature difference method under controlled vapor pressure, can be explained by the equality of the arsenic chemical potentials, and the dominating point defects are arsenic interstitial atoms and arsenic vacancies, but not gallium vacancies and gallium interstitials.

On the basis of this theory, we will present models for atomic diffusion of impurities and point defects. We discuss sulfur diffusion, self-diffusion of gallium and arsenic, and silicon diffusion. they can well explain the arsenic vapor pressure dependences, and the comparison with known experiments gives reasonable values for formation energies and migration energies. We also discuss impurity-enhanced diffusion at a heterostructure interface where sharp discontinuities of gallium and aluminum chemical potentials are present.

Key words. chemical potentials, interstitials and vacancies

AMS(MOS) subject classifications.

1. Introduction.

The deviation from stoichiometry was found to greatly change the material properties. In the case of $GaAs$, the heat-treatment experiment under the applied arsenic pressure showed the existence of the exact stoichiometric vapor pressure, and it was found that interstitial arsenic atoms and arsenic vacancies were dominant point defects governing the deviation from stoichiometry [1-3]. On the other hand, the liquid phase and melt growth experiments were made with applied arsenic vapor pressure upon the solution [3-8], and it was found that the deviation from the stoichiometry was controlled by the applied vapor pressure and the stoichiometric crystals were segregated at just the same arsenic vapor pressure as that of the heat-treatment. The grown crystals were very perfect. This growth method was called the temperature difference method under controlled vapor pressure ($TDM \cdot CVP$). Figure 1 illustrates the experimental methods of the heat-treatment and $TDM \cdot CVP$ growth.

The stoichiometry-controlled crystal growth in $TDM \cdot CVP$ was found to be due to the increased saturation solubility of the solution under applied vapor pressure [6,7] and it was found that the equality of the chemical potentials of arsenic holds between the three phases. The details of the experiments and the chemical potential approach can be referred to Reference [7].

Atomic diffusions in $GaAs$ are also thought to be greatly influenced by the deviation from stoichiometry. However, most of the discussions so far have not given strong attentions on this point. In this paper, we discuss the atomic diffusions

---
*Department of Materials Science, Faculty of Engineering, Tohoku University, Aoba Aramaki, Aoba-ku, Sendai 980, JAPAN.
**Tohoku University, 2-1-1 Katahira, Aoba-ku, Sendai 980, JAPAN.

on the basis of the chemical potential approach developed for $TDM \cdot CVP$. We deal with diffusion of sulfur, self diffusion of arsenic and gallium, and also diffusion of silicon. As we show in the following sections, these are typical of interstitial diffusion, diffusion via arsenic vacancies, and diffusion with site transfer. In these discussions we assume that corresponding point defects are in thermal equilibrium under the applied vapor pressure. In the final section, however, we deal with atomic diffusion at a hetero-interface, which is an example in which equality of the chemical potentials does not hold, which causes characteristic interface mixing phenomena.

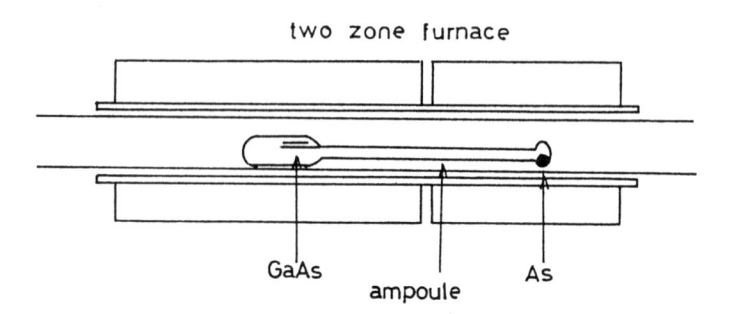

Figure 1a. Heat-treatment of $GaAs$ under controlled arsenic vapor pressure.

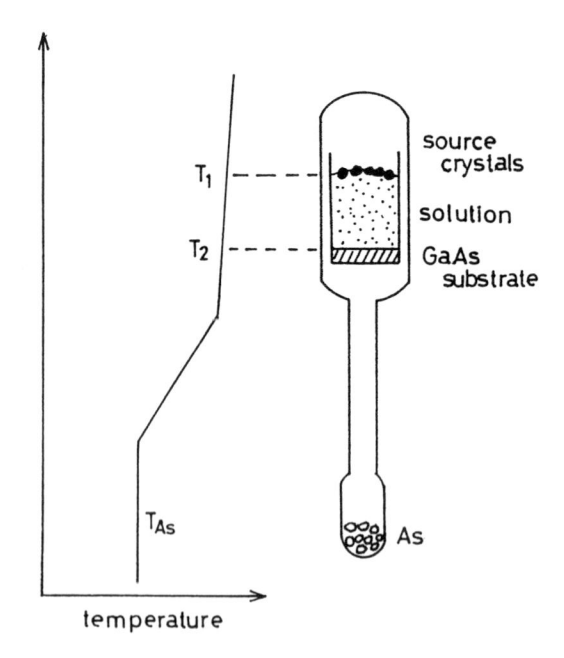

Figure 1b. Schematic diagram of stoichiometry controlled crystal growth method ($TDM \cdot CVP$).

2. The stoichiometry control by the applied arsenic vapor pressure.

In the $TDM \cdot CVP$ crystal growth and also in the heat-treatment of $GaAs$ under applied arsenic vapor pressure, the deviation from the stoichiometry of the crystal is determined by the equality of the arsenic chemical potentials. That is,

(2.1) $\qquad \mu_{As}^g = \mu_{As}^l = \mu_{As}^s \quad$ (for $TDM \cdot CVP$)

(2.2) $\qquad \mu_{As}^g = \mu_{As}^s \qquad$ (for heat-treatment).

There are four kinds of point defects which possibly cause the deviation from stoichiometry, except anti-lattices. They are an arsenic interstitial atom, I_{As}, arsenic vacancy. V_{As}, gallium interstitial, I_{Ga}, and gallium vacancy V_{Ga}. However, we have experimentally shown that I_{As} and V_{As} are dominant, while I_{Ga} and V_{Ga} are much less in concentration. In such a case, the As_4 vapor pressure which gives the exact stoichiometry (which we call the optimum vapor pressure in $TDM \cdot CVP$) is determined by the equality of the concentration of I_{As} and V_{As}. Figure 2 shows the experimental stoichiometric vapor pressures in $TDM \cdot CVP$ and in the heat-treatment, together with the calculation based on I_{As} and V_{As}. These three curves are in a very good agreement. Table 1 gives the free energies of formation for I_{As} and V_{As} adopted for the calculation, which were determined from the change in the lattice parameter. Recently, photocapacitance studies have shown [9-11] that the formation energy of the interstitial arsenic atoms is $\Delta H_{I_{As}}^{F'} = 1.1 eV$, which is just the same as adopted in the calculation. From these results, the formation energies of I_{As} and V_{As} listed in Table 1 are thought to be reliable enough.

	ΔH^F	$\Delta H^{F'}$	ΔS_V
I_{As}	1.52 eV	1.19 eV (1.1 eV[*])	10.48 e.u.
V_{As}	1.38 eV	1.71 eV	1.18 e.u.

[*]Reference [10]

Table 1. Free energy parameters for arsenic interstitial atoms and arsenic vacancies in $GaAs$.

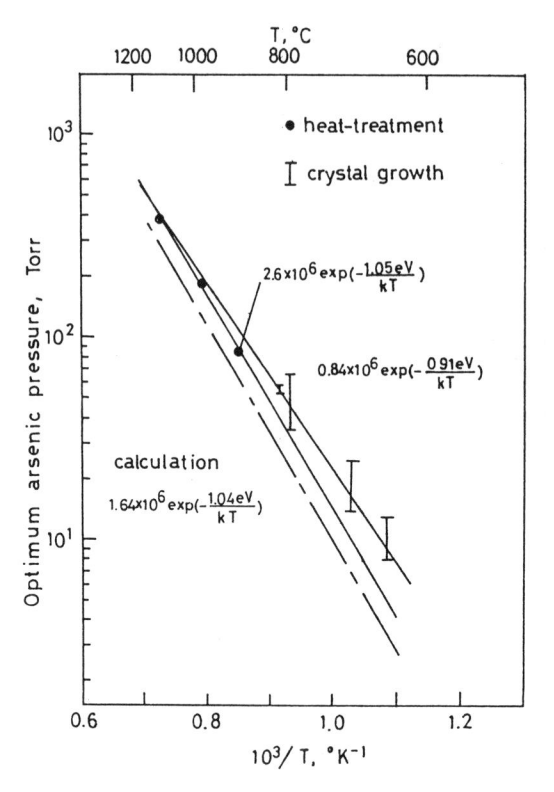

Figure 2. Optimum arsenic vapor pressure as a function of temperature

We consider the following reactions for the formation of I_{As} and V_{As}.

(2.3) $\qquad As$ (surface, solid) $= I_{As}$; ΔG^F_{IAs},
(2.4) $\qquad As$ (lattice site) $= V_{As} + As$ (surface, solid) ; ΔG^F_{VAs}.

ΔG^F_{IAs} and ΔG^F_{VAs} are the free energy differences per a molecule, and described as

(2.5)
$$\Delta G^F_{IAs} = \Delta H^F_{IAs} - T\Delta S^F_{IAs}$$
$$\Delta G^F_{VAs} = \Delta H^F_{VAs} - T\Delta S^F_{VAs}$$

where ΔH^F_{IAs} and ΔH^F_{VAs} are the entalpies, and ΔS^F_{IAs} and ΔS^F_{VAs} are the entropies of vibration.

In order to refer to the applied As_4 vapor pressure, we need the following reaction equation.

(2.6) $\qquad As$ (surface, solid) $= 1/4 As_4$; $\Delta G^{\text{sub}}_{As}$

where ΔG_{As}^{sub} is the free energy of sublimation of arsenic element. If we write the reactions of formation of I_{As} and V_{As} as,

(2.7) $$\frac{1}{4}As_4 \text{ (gas)} = I_{As}; \quad \Delta G_{IAs}^{F'}$$

(2.8) $$As \text{ (lattice site)} = V_{As} + \frac{1}{4}As_4 \text{ (gas)}; \quad \Delta G_{VAs}^{F'}.$$

We have the following relations from Equations (2.3) to (2.8).

(2.9) $$\Delta G_{IAs}^{F'} = \Delta G_{IAs}^{F} - \Delta G_{As}^{sub}$$
$$\Delta G_{VAs}^{F'} = \Delta G_{IAs}^{F} + \Delta G_{As}^{sub}.$$

In these expressions F' and F mean the free energies referring to the solid and gas phase arsenic, respectively. When the chemical potential of the gas phase is equal to that of the solid phase, we have the following equations from Equations (2.7) and (2.8).

(2.10) $$\frac{[I_{As}]}{(P_{As_4})^{\frac{1}{4}}} = \exp\left(-\frac{\Delta G_{IAs}^{F'}}{kT}\right)$$

(2.11) $$[V_{As}](P_{As_4})^{\frac{1}{4}} = \exp\left(-\frac{\Delta G_{VAs}^{F'}}{kT}\right).$$

In these expressions, $[I_{As}]$ and $[V_{As}]$ are defined as $[I_{As}] = N_{IAs}/N_{IAs}^0$ and $[V_{As}] = N_{VAs}/N_{SAs}^0$, where N_{IAs} and N_{VAs} are concentrations of I_{As} and V_{As}, and N_{IAs}^0 and N_{SAs}^0 are the concentrations of the whole available interstitial and the substitutional arsenic sites.

Assuming that $N_{IAs}^0 = N_{SAs}^0$, the As_4 vapor pressure corresponding to the exact stoichiometry, which we call the optimum vapor pressure P_{As}^{opt}, can be obtained from the condition that $[I_{As}] = [V_{As}]$.

That is, from Equations (2.10) and (2.11) we have

(2.12) $$P_{As_4}^{opt} = \exp\left(-\frac{4\Delta G_{As}^{sub}}{kT}\right) \exp\left\{-\frac{2(\Delta G_{VAs}^{F} - \Delta G_{IAs}^{F})}{kT}\right\}$$
$$= \exp\left\{-\frac{2(\Delta G_{VAs}^{F'} - \Delta G_{IAs}^{F'})}{kT}\right\}.$$

Figure 2 has shown the calculation by this equation with the parameters listed in Table 1. As for the gallium vacancies, V_{Ga}, and gallium interstitials, I_{Ga}, the similar expressions are possible, though their concentrations are thought to be much smaller.

That is, the formation energies are defined from the following reactions,

(2.13) $\quad Ga \text{ (surface, solid)} = I_{Ga}; \quad \Delta G_{IGa}^{F}$,

(2.14) $\quad Ga \text{ (lattice site)} = V_{Ga} + Ga \text{ (surface, solid)}; \quad \Delta G_{VGa}^{F}.$

In order to refer to the As_4 vapor pressure, we use the following equation

(2.15) $\qquad GaAs$ (solid) $= Ga$ (solid) $+ 1/4As_4$ (gas) ; ΔG^{sub}_{GaAs}

where ΔG^{sub}_{GaAs} is the free energy of the sublimation of $GaAs$ solid phase. Then, we have

(2.16) $\qquad GaAs$ (solid) $= I_{Ga} + \frac{1}{4}As_4$ (gas) ; $\Delta G^{F'}_{IGa}$

(2.17) $\qquad \frac{1}{4}As_4 = V_{Ga} + As$ (lattice site) ; $\Delta G^{F'}_{VGa}$

with

(2.18) $\qquad \begin{aligned} \Delta G^{F'}_{IGa} &= \Delta G^{F}_{IGa} + \Delta G^{sub}_{GaAs} \\ \Delta G^{F'}_{VGa} &= \Delta G^{F}_{VGa} - \Delta G^{sub}_{GaAs} \end{aligned}$

Equations (2.16) and (2.17) gives the arsenic vapor pressure and temperature dependences of $[I_{Ga}]$ and $[V_{Ga}]$ as follows.

(2.19) $\qquad [I_{Ga}](P_{As_4})^{\frac{1}{4}} = \exp\left(-\frac{\Delta G^{F'}_{IGa}}{kT}\right)$

(2.20) $\qquad [V_{Ga}](P_{As_4})^{-\frac{1}{4}} = \exp\left(-\frac{\Delta G^{F'}_{VGa}}{kT}\right).$

3. Diffusion of sulfur.

There is a detailed experiment on the diffusion of sulfur in $GaAs$ under applied arsenic vapor pressures by Young and Pearson [12]. The diffusion coefficient D increases as the square root of the As_4 vapor pressure, but it saturates at a higher vapor pressure as shown in Figure 3. They proposed that the complex of gallium divacancy and sulfur donor is responsible for the diffusion. However, as pointed out in Sec. 2, the equilibrium concentrating of the gallium divacancies should be so small that they could not be dominant migrating species. On the other hand, B. Tuck assumed the arsenic vacancies in his calculation of the diffusion profile [13]. But we think that arsenic vacancies cannot explain the vapor pressure dependence of D, in spite of his assertion.

In place of them, we propose an interstitial sulfur diffusion model. First, it should be pointed out that the cross over-points of the square root line and the constant D line, is close to the optimum vapor pressure described in Section 2, both at $T = 1000°C$ and $1130°C$, as denoted by arrows in Figure 3. The model must explain this fact, as well as the vapor pressure dependence.

We assume that a substitutional sulfur donor S^+_{As} changes to an interstitial sulfur, I^+_s, and an arsenic vacancy, V^0_{As}, or we assume that, in the presence of an interstitial arsenic atom I^0_{As}, S^+_{As} changes to an interstitial molecular complex $(Is \cdot I_{As})^+$ and an arsenic vacancy.

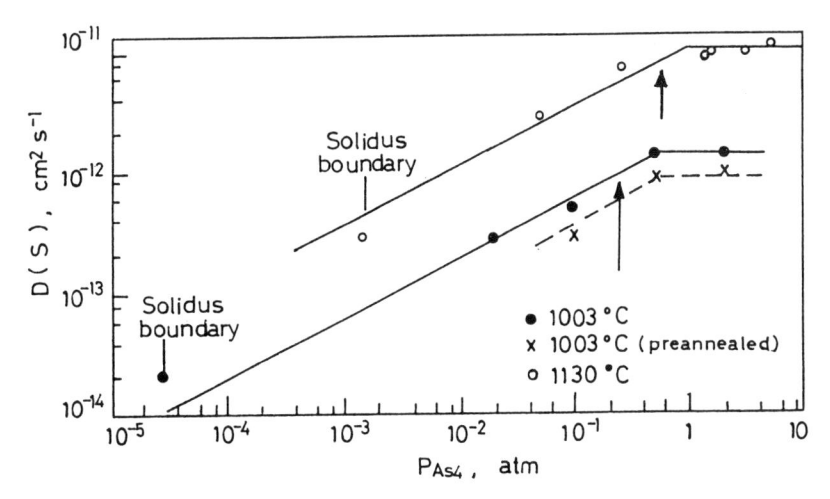

Figure 3. Diffusion coefficient of sulfur in $GaAs$ by Young and Pearson [12]. Arrows indicate the optimum arsenic vapor pressures by Nishizawa, Okuno and Tadano [5].

The following reaction equations should hold in the above two cases, respectively.

(3.1) $$S^+_{As} \rightleftarrows I^+_s + V^0_{As}; \quad \Delta Gi_1$$
(3.2) $$S^+_{As} + I^0_{As} \rightleftarrows (I_s \cdot I_{As})^+ + V^0_{As}; \quad \Delta Gi_2.$$

We have assumed that the charge states of sulfur do not change in this reaction because there is no such an observation of strong concentration dependence as in Z_n diffusion [17]. Also, charge states of arsenic vacancies and arsenic interstitial atoms have been assumed to be neutral as was described in Section 2.

For the interstitial diffusion under thermal equilibrium of V_{As} and I_{As}, the former equation gives the $(P_{As_4})^{\frac{1}{4}}$ dependence of D, while the latter gives $(P_{As_4})^{\frac{2}{4}}$ dependence. Because of the vapor pressure dependence and also because the diffusion coefficient of sulfur is much smaller than that of zinc for which it is well established that zinc interstitial atoms have a form of isolated atoms [14], it will be reasonable to assume that an interstitial sulfur and an interstitial arsenic atom make a complex like a molecule as described by Equation (3.2).

Following the diffusion equation which was established in the case of zinc diffusion, the diffusion equation for sulfur is given by

(3.3) $$\frac{\partial}{\partial t}(N_{\text{sub}} + N_{\text{inter}}) = \frac{\partial}{\partial x}\left(D'_s \frac{\partial N_{\text{sub}}}{\partial x} + D'_i \frac{\partial N_{\text{inter}}}{\partial x}\right)$$

where N_{sub} and N_{inter} are concentrations of substitutional and interstitial species and D'_s and D'_i are their intrinsic diffusion coefficients, respectively. If we can assume that $N_{sub} \gg N_{inter}$, and the reaction given by Equation (3.2) under the applied arsenic vapor pressure is fast enough, we get

$$\frac{\partial}{\partial t} N_{sub} = \frac{\partial}{\partial x} \left(D'_{sub} + D'_{inter} \frac{\partial N_{inter}}{\partial N_{sub}} \right) \frac{\partial N_{sub}}{\partial x} \tag{3.4}$$

which gives the following effective diffusion coefficients, D_{sub}, for substitutional sulfur

$$D_{sub} = D'_{sub} + D'_{inter} \frac{\partial N_{inter}}{\partial N_{sub}}. \tag{3.5}$$

As will be discussed in the next section, the substitutional impurity diffusions via vacancies which are typical in elemental semiconductors are thought to be hard to occur in the covalent compound semiconductors. Therefore, we consider only the second term, i.e.,

$$D_{sub} \simeq D'_{inter} \frac{\partial N_{inter}}{\partial N_{sub}}. \tag{3.6}$$

Under the arsenic chemical potential given by the applied arsenic vapor pressure, the equilibrium for the equation (3.2) gives the following equation.

$$\frac{[I_s \cdot I^+_{As}] \cdot [V^0_{As}]}{[S^+_{As}] \cdot [I^0_{As}]} = \exp\left(-\frac{\Delta G_{i2}}{kT}\right) \tag{3.7}$$

where $[S^+_{As}]$ is the concentration of S^+_{As}, relative to the concentration of the whole available sites for S^+_{As}, and all other notations $[I^0_{As}]$, $[V^0_{As}]$ and $[I_{As} \cdot I^+_{As}]$ are also relative concentrations of each species.

Assuming that the densities of the whole available sites for S^+_{As} and $(I_s \cdot I_{As})^+$ are the same, we get

$$\frac{\partial N_{inter}}{\partial N_{sub}} = \frac{[I^0_{As}]}{[V^0_{As}]} \exp\left(-\frac{\Delta G_{i2}}{kT}\right) \tag{3.8}$$

so that the effective diffusion constant for substitutional sulfur is given by

$$D_{sub} = D'_{inter} \frac{[I^0_{As}]}{[V^0_{As}]} \exp\left(-\frac{\Delta G_{i2}}{kT}\right). \tag{3.9}$$

If the thermal equilibrium is established under the applied As_4 vapor pressure, $[I^0_{As}]$ and $[V^0_{As}]$ are given by Equations (2.10) and (2.11) in section 2, respectively.

Then, D_{sub} is expressed as follows using P_{As_4} and the free energies of formation for I_{As}, V_{As} and $(I_s \cdot I_{As})^+$.

$$D_{sub} = D'_{inter} (P_{As_4})^{\frac{2}{4}} \exp\left(-\frac{\Delta G^{F'}_{IAs} - \Delta G^{F'}_{VAs} + \Delta G_{i2}}{kT}\right). \tag{3.10}$$

On the other hand, the stoichiometric vapor pressure, $P_{As_4}^{opt}$ is given by Equation (2.12), at which $[I_{As}] = [V_{As}]$ holds. Therefore, the diffusion constant at the stoichiometric vapor pressure, D_{sub}^{opt} is given by

$$D_{sub}^{opt} = D_{inter}' \exp\left(-\frac{\Delta G_{i2}}{kT}\right). \tag{3.11}$$

D_{inter}' is approximately given by the following form using the free energy of migration of the interstitial molecule $\Delta G_{m2} = \Delta H_{m2} - T\Delta S_{m2}$

$$D_{inter}' \simeq \frac{1}{6} d^2 \bar{\nu} \exp\frac{\Delta S_{m2}}{k} \exp\left(-\frac{\Delta H_{m2}}{kT}\right) \tag{3.12}$$

where d is a distance corresponding to a jump for interstitial migration, and $\bar{\nu}$ is a jump frequency.

Therefore, the activation energy Q_s^{opt} for the sulfur diffusion under the stoichiometric vapor pressure $p_{As_4}^{opt}$ is given by

$$Q_s^{opt} = \Delta H_{m2} + \Delta H_{i2}$$

where ΔH_{i2} is the entalpy part of ΔG_{i2}.

Young and Pearson gave $Q_s = 2.6eV$, in the region where the vapor pressure dependence of the diffusion coefficient saturates, which should be close to Q_s^{opt} as was shown in Figure 3. As a rough estimation of ΔH_{m2} and ΔH_{i2}, the migration energies of self-interstitial diffusion in silicon and germanium was estimated to be very small, in the range of $0 \sim 0.3eV$ [15,16]. In the case of interstitial molecule, ΔH_{m2} should be considerably larger than $0.3eV$.

There are two important interstitial sites, which have tetragonal and hexagonal symmetries as illustrated in Figure 4. In elemental semiconductors, Si and Ge, it is considered that the tetragonal site has a lower energy, so that the free energy for migration corresponds to a jump through the hexagonal interstitial site [19]. Therefore, as a possible model of interstitial complex, two atoms S and As are in the neighboring tetragonal interstitial sites as shown in Figure 4, and they can move as a whole, with each atoms moving through a hexagonal saddle point, to one of four equivalent neighbouring locations. Therefore, ΔH_{m2} for a molecule should not be so much larger than twice of ΔH_m for isolated atoms, that is, we roughly estimate that $\Delta H_{m2} \gtrsim 0.6eV$.

As for ΔH_{i2}, it should be smaller than the activation energy for Frenkel pair of I_{As} and V_{As}, that is

$$\text{As (lattice)} \rightleftarrows I_{As} + V_{As}; \Delta G_{As}^{Frenkel} = G_{I_{As}}^F + \Delta G_{V_{As}}^F. \tag{3.13}$$

From Table 1, $\Delta H_{As}^{Frenkel} = 2.9eV$ is obtained and we get $\Delta H_{i2} < \Delta H_{i1} < 2.9eV$.

Therefore, it is understood that $Q_s^{opt} = \Delta H_{m2} + \Delta H_{i2} = 2.6eV$ is in the reasonable range.

Comparing with the present sulfur case, it was assumed in the case of Zn diffusion that zinc interstitials are isolated atoms. If the attractive interaction of zinc and arsenic interstitial atoms is too strong, a stable complex may not be formed because they tend to find a place in a single interstitial cell so that they destroy a lattice and one of them return to a substitutional site. We assume that the molecular complex $(I_s \cdot I_{As})^+$ is stable because of a weaker interaction between I_s and I_{As} as illustrated in Figure 4.

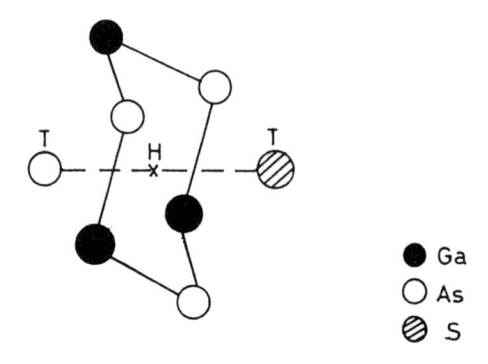

Figure 4. Interstitial sites in the $GaAs$ lattice, and a model of a comlex of sulfur and arsenic interstitial atoms.

3.1 Saturation of the diffusion coefficient of sulfur at a high arsenic vapor pressure.

It was shown that the density of the stacking faults in $GaAs$ grown by $TDM\ CVP$ rapidly increases with increasing the arsenic vapor pressure when it exceeds the optimum vapor pressure [3]. This fact suggests that arsenic interstitial atoms tend to aggregate each other to form larger scale defects like dislocations, stacking faults, and precipitates at higher vapor pressures exceeding the optimum vapor pressure [3,7].

We assume here that, at a high arsenic vapor pressure, irreversible aggregation reaction takes place while the point defects like arsenic interstitial atoms and arsenic vacancies are still nearly in equilibrium with the applied vapor pressure. If we can assume an aggregation reaction takes place when an interstitial molecule find an arsenic interstitial atom, the reaction can be expressed as,

$$(3.14) \qquad (I_s \cdot I_{As})^+ + I_{As}^0 \xrightarrow{k} \{2As + S\}$$

where k is the reaction rate, and $\{2As + S\}$ denotes the smallest aggregated configuration. Then, we have the following rate equation,

$$(3.15) \qquad \frac{d}{d_t}[I_s \cdot I_{As}^+] = -k[I_s \cdot I_{As}^+] \cdot [I_{As}^0].$$

Also, the reaction described by Equation (3.2) can be expressed by using the rate constants k_1 and k_2 as follows

$$S^+_{As} + I^0_{As} \underset{k_2}{\overset{k_1}{\rightleftarrows}} (I_s \cdot I_{As})^+ + V^0_{As} : \Delta G_{i2}$$

(3.16)
$$\frac{d}{dt}[I_s \cdot I^+_{As}] = k_1[S^+_{As}][I^0_{As}] - k_2[I_s \cdot I^+_{As}][V^0_{As}]$$

$$\text{with } \frac{k_1}{k_2} = \exp\left(-\frac{\Delta G_{i2}}{kT}\right).$$

In a steady state, the whole generation rate of $(I_s \cdot I_{As})^+$ equals 0, i.e.,

(3.17) $\quad \frac{d}{dt}[I_s \cdot I^+_{As}] = k_1[S^+_{As}][I^0_{As}] - k_2[I_s \cdot I^+_{As}][V^0_{As}] - k[I_s \cdot I^+_{As}] \cdot [I^0_{As}] = 0.$

Then, we have

(3.18) $\quad \dfrac{\partial N_{\text{inter}}}{\partial N_{\text{sub}}} = \dfrac{[I_s \cdot I^+_{As}]}{[S^+_{As}]} = \dfrac{1}{\frac{[V^0_{As}]}{[I^0_{As}]}\exp\left(\frac{\Delta G_{i2}}{kT}\right) + \frac{k}{k_1}}.$

Therefore, the equivalent diffusion coefficient is given by

(3.19) $\quad D_{\text{sub}} = D'_{\text{inter}} \dfrac{1}{\frac{1}{\left(\frac{\partial N_{\text{inter}}}{\partial N_{\text{sub}}}\right)_0} + \frac{k}{k_1}}.$

with

$$\left(\frac{\partial N_{\text{inter}}}{\partial N_{\text{sub}}}\right)_0 = \frac{[I^0_{As}]}{[V^0_{As}]}\exp\left(-\frac{\Delta G_{i2}}{kT}\right) = (P^{\frac{1}{4}}_{As_4})^2 \exp\left(-\frac{\Delta G^{F'}_{IAs} - \Delta G^{F'}_{VAs} + \Delta G_{i2}}{kT}\right).$$

$\left(\frac{\partial N_{\text{inter}}}{\partial N_{\text{inter}}}\right)_0$ is just the same as was given by Equation (3.8).

The diffusion coefficient in the saturation region is given by $D^{\text{sat}}_{\text{sub}} = D'_{\text{inter}} \frac{k_1}{k}$, and the critical vapor pressure is given by

(3.20) $\quad P^{\text{critical}}_{As_4} = \left(\frac{k_1}{k}\right)^2 \exp\left\{\frac{2(\Delta G^{F'}_{IAs} - \Delta G^{F'}_{VAs} + \Delta G_{i2})}{kT}\right\} = \left(\frac{k_2}{k}\right)^2 p^{\text{opt}}_{As_4}.$

The experimental fact that the $P^{\text{critical}}_{As_4}$ is close to the optimum vapor pressure means that the values of k and k_2 are not greatly different each other.

Comparing the two reactions,

(3.16) $\quad (I_s \cdot I_{As})^+ + V^0_{As} \overset{k_2}{\rightleftarrows} S^+_{As} + I^0_{As}$

(3.14) $\quad (I_s \cdot I_{As})^+ + I^0_{As} \overset{k}{\to} \{2As + S\}.$

The reaction rate constants k_2 and k are determined by the random walk areas of $(I_s \cdot I_{As})^+$, V^0_{As} and I^0_{As}, as far as we assume they immediately react when they come to the nearest neighbour sites each other. Therefore, if interstitial arsenic atoms are isolated each other, k may be larger than k_2 because the diffusion constant of I_{As} is thought to be the largest. However, if interstitial arsenic atoms are interacting each other at such a high concentration that aggregation takes place and effective diffusion constant of I_{As}^0 is reduced, then the diffusion constant of $(I_s \cdot I_{As})^+$ determines both k_2 and k so that $k_2 \simeq k$ will hold. The fact that the cross over point is close to the optimum vapor pressure means that the above mentioned mechanism is essentially true. However, more precisely, the cross over point is about twice times the optimum pressure and the curve bends more sharply than Equation (3.19) predicts. We think, therefore, that actual aggregation takes place more suddenly when the concentration of arsenic interstitial atoms exceeds some level.

Finally in this section, Figure 5 shows the calculated temperature dependence of the diffusion coefficient of sulfur at a stoichiometry and at a gallium rich liquidus line as well as cross over point with assuming $\Delta H_{m2} + \Delta H_{i2} = 2.6 eV$, and $\left(\frac{k_2}{k}\right)^2 = 2$.

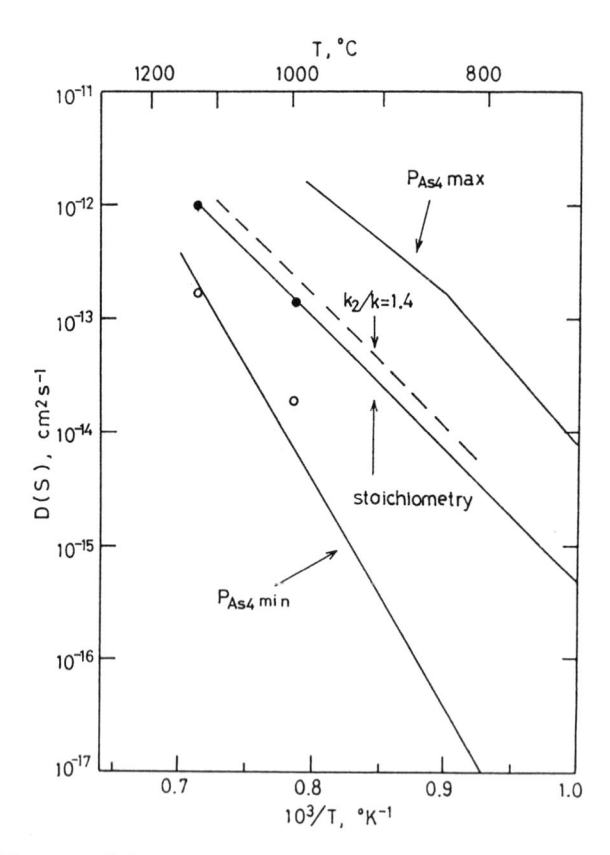

Figure 5. Calculated temperature dependence of the diffusion coefficient of sulfur in $GaAs$, together with experimental points from Young and Pearson (• at the stoichiometric vapor pressure, o at the solidus boundary).

4. Self-diffusion of Ga and As.

Although there is an early work on a self-diffusion by Goldstein [17], the arsenic vapor pressure was not controlled. Recently, Kashiwagi [18] has made a more precise experiment of the self-diffusion of Ga and As under the applied arsenic vapor pressures. His result is greatly different from that of Goldstein as for the activation energies.

Kashiwagi's result is that both D_{As} and D_{Ga} is proportional to $(P_{As_4})^{-\frac{1}{4}}$, and the highest values near to the gallium-rich liquidus line have been given as a function of temperature as follows

$$D_{As} = 2.3 \times 10^{-5} \exp\left(-\frac{2.06eV}{kT}\right)$$
$$D_{Ga} = 5.2 \times 10^{-5} \exp\left(-\frac{2.14eV}{kT}\right).$$

The most striking result is that D_{As} and D_{Ga} are nearly equal. Although the above expressions are a little bit different from each other, they are equal within the experimental uncertainty over wide ranges of vapor pressure and temperature.

It should be first pointed out that the diffusants are isotopic As and Ga (we describe them As' and Ga') while the arsenic vapor pressure is that of usual natural arsenic atoms. As a result, the chemical potential for isotopic arsenic, μ'_{As}, is without control, and $[I'_{As}]$ should be much smaller than $[I_{As}]$. On the other hand, there is no discrimination between As' and As for V_{As}. Therefore, this kind of experiment is not related to the diffusion of arsenic interstitial atoms.

We present a model in which both arsenic and gallium atoms diffuse assisted by arsenic vacancies.

It has been usually assumed that V_{As} and V_{Ga} can migrate like shown in Figure 6 (a), that is, an arsenic atom jumps to V_{As} directly from one of the next nearest lattice sites. As a comparison, in silicon and other elemental semiconductors, an atom need only to jump to the nearest neighbour lattice site, as shown in Figure 6 (b). The latter need only stretching of the lattice bonds as illustrated by dotted lines, but in the former case (a) the bonds should be broken and the atom must go through an interstitial site. It will need a much higher migration energy than in the case of (b). On the other hand, if an atom in compound semiconductors could jump to a vacancy at the nearest neighbour lattice site like in elemental semiconductors, then, a line of antilattices would be generated as illustrated in Figure 6 (c). From this discussion, it is understood that a simple diffusion mechanism via vacancies as in silicon is hard to be considered in covalent compound semiconductors.

The experimental fact that $D_{Ga} = D_{As}$ should not be accidental but it implies that Ga' (or Ga) and As' (or As) jump as a pair in the presence of an arsenic vacancy.

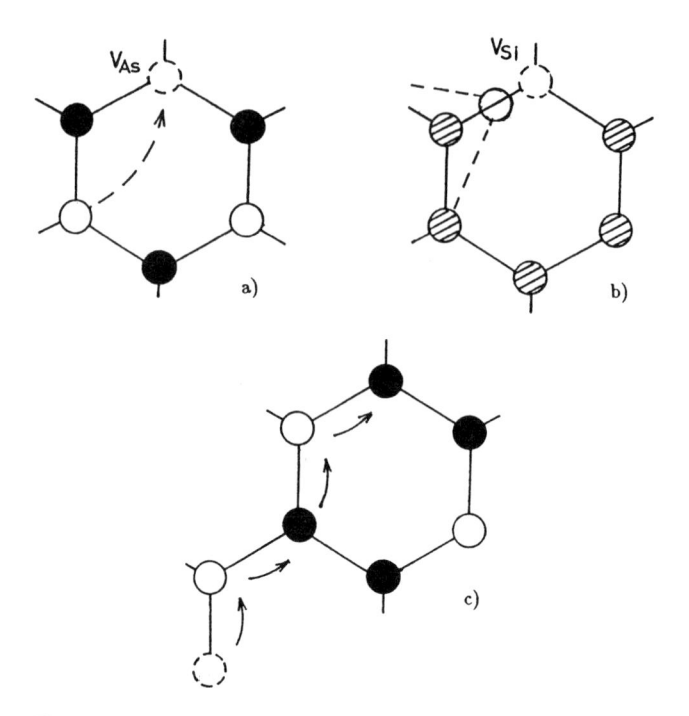

Figure 6. Illustrations of vacancy migration: a) a simple migration model in $GaAs$, b) migration in Si, c) migration forming antilattices.

Figure 7 illustrates a possible migration process. First, a Ga' (or Ga) jumps to the nearest neighbour V_{As} site and, at the same time, As' (or As) nearest to the Ga' jumps to the former Ga' site, so that the V_{As} moves to the next nearest lattice site and paired antilattices $Ga' - As'$ are formed. This paired antilattices in Figure 7 (b) do not mean a stable energy state but a saddle point through which $(Ga' - As')$ move to the final stable state. The latter half of the step is as follows. The three atoms of the paired antilattices $(Ga' - As')$ and a neighbouring Ga denoted Ga'' in Figure 7 (c) cause interchanges between them as shown by the three arrows in Figure 7 (c) and relax to a final state shown in Figure 7 (d). It is assumed that the saddle point energy of the paired antilattices $(Ga' - As')$ is enough high to cause the movement of the third atom Ga''. As for the possibility of an interchange within $(Ga' - As')$, it will need a higher energy than the three body interchange.

The migration of an arsenic vacancy should be much easier in this model than that in the simpler process illustrated in Figure 6 (a). There may be a similar process based on $V_{Ga'}$ but we can assume that the concentration of V_{Ga} is much smaller than that of V_{As} as explained in Section 2. Therefore, the diffusion constant of As and Ga are the same, and given by the following form in terms of $[V_{As}]$.

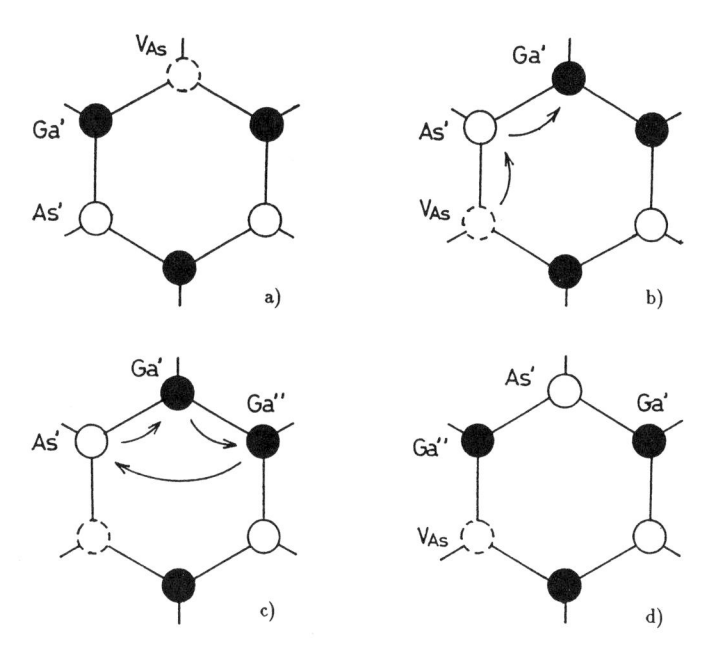

Figure 7. Model of gallium and arsenic self-diffusion in $GaAs$ via arsenic vacancies. $Ga'-As'$ in a) are finally transferred as $Ga'-As'$ in d).

(4.1) $$D_{As} = D_{Ga} \fallingdotseq \frac{1}{6}d^2\nu[V_{As}]\exp\left(-\frac{\Delta G^m_{\text{pair}}}{kT}\right)$$

where $\Delta G^m_{\text{pair}} = \Delta H^m_{\text{pair}} - T\Delta S^m_{\text{pair}}$ is the free energy of the saddle point corresponding to paired antilattices with V_{As} at the nearest site.

Using Equation (2.10) the diffusion coefficient is expressed in terms of the applied As_4 vapor pressure and temperature as follows

(4.2) $$D_{As} = D_{Ga} \fallingdotseq \frac{1}{6}d^2\nu(P_{As_4})^{-\frac{1}{4}}\exp\left(-\frac{\Delta G^m_{\text{pair}} + \Delta G^{F'}_{VAs}}{kT}\right).$$

In order to compare with the experiment, we must know the expression at the gallium rich liquidus line at which $[V_{As}]$ becomes maximum and P_{As_4} becomes minimum.

$(P_{As_4}^{-\frac{1}{4}})$ min is obtained from the following equation

(4.3) $$GaAs \text{ (lattice)} = Ga \text{ (liquid)} + \frac{1}{4}As_4; \quad \Delta^{\text{sub}'}_{GaAs}.$$

This is composed from the following equations

(2.15) $\quad\quad GaAs \text{ (solid)} = Ga \text{ (solid)} + \frac{1}{4}As_4; \quad \Delta G_{GaAs}^{sub}$

(4.4) $\quad\quad Ga \text{ (solid)} = Ga \text{ (liquid)} ; \Delta G_{Ga}^{f}$

so that we obtain

(4.5) $\quad\quad \Delta G_{GaAs}^{sub'} = \Delta G_{GaAs}^{sub} + \Delta G_{Ga}^{f}$

where ΔG_{GaAs}^{sub} is the free energy of sublimation of $GaAs$ and ΔG_{Ga}^{f} is the free energy of fusion of gallium. Equation (4.3) gives

(4.6) $\quad\quad a_{Ga}^{\ell} \cdot (P_{As_4})^{\frac{1}{4}} = \exp\left(-\frac{\Delta G_{GaAs}^{sub'}}{kT}\right).$

where a_{Ga}^{ℓ} is the activity of liquid gallium.

At the gallium rich liquidus line at temperature far below the melting point, the solubility of arsenic in liquid gallium is small and we can assume $a_{Ga} \simeq 1$.

Then, we have

(4.7) $\quad\quad (P_{As_4}^{\frac{1}{4}})\min \simeq \exp\left(-\frac{\Delta G_{GaAs}^{sub'}}{kT}\right).$

Therefore, the highest value of the diffusion coefficient of As and Ga is given by

(4.8) $\quad D_{As}^{\max} = D_{Ga}^{\max} \simeq \frac{1}{6}d^2\nu \exp\left(-\frac{\Delta G_{pair}^{m} + \Delta G_{VAs}^{F'} - \Delta G_{GaAs}^{sub'}}{kT}\right).$

In the expression $D = D_0 \exp\left(-\frac{Q}{kT}\right)$, $Q = 2.1eV$ was obtained by Kashiwagi, that is, we get $Q = \Delta H_{pair}^{m} + \Delta H_{VAs}^{F'} - \Delta H_{GaAs}^{sub'} \simeq 2.1eV$. Using the known value $\Delta H_{GaAs}^{sub} = 1.14eV$, $\Delta H_{Ga}^{f} = 0.058eV$ we get $\Delta H_{GaAs}^{sub'} = 1.20eV$. Also, we have shown that $\Delta H_{VAs}^{F'} = 1.71eV$ in Table 1. Therefore, we get $\Delta H_{pair}^{m} = 1.6eV$.

This corresponds to the energy height of the saddle point, and it means the activation energy of migration of V_{As}.

The migration energies of vacancies in silicon and germanium have been estimated to be $1.06 \sim 1.09eV$, and $0.95 \sim 0.98eV$, respectively. Considering that the state in Figure 7 (b) is higher in energy than the saddle point state in elemental semiconductors shown by the dotted line in Figure 6 (b), the estimated value of ΔH_{pair}^{m} should be thought to be in a reasonable range. Using the values $\Delta H_{pair}^{m} = 1.6eV$ and $\Delta H_{VAs}^{F'} = 1.71eV$, the self-diffusion coefficients in Equation (4.2) are described as follows.

(4.9) $\quad\quad D_{As} = D_{Ga} = D_0(P_{As_4})^{-\frac{1}{4}} \exp\left(-\frac{3.3eV}{kT}\right).$

Figure 8 shows the calculated temperature and vapor pressure dependence in comparison with the experiments by Kashiwagi and Goldstein.

Considering that Goldstein's experiment was made under excess arsenic vapor pressures but without control, his experimental points are almost within a calculated region, but the activation energies may be meaningless.

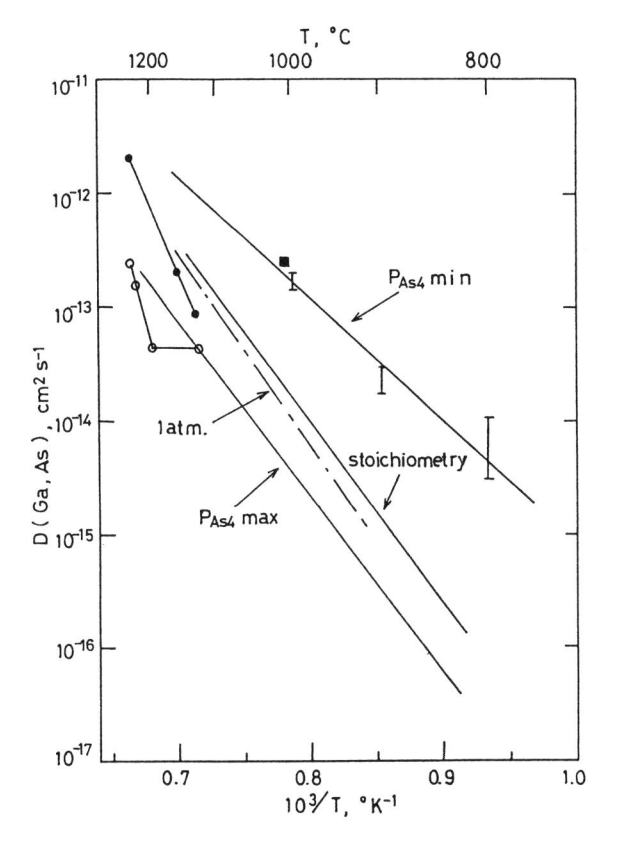

Figure 8. Calculated temperature dependence of the self-diffusion coefficients of gallium and arsenic in $GaAs$, $D(Ga) = D(As)$. I, experimental data from Kashiwagi [18]. • and ○ are $D\ (Ga)$ and $D\ (As)$, respectively, by Goldstein [17].

5. Diffusion of silicon.

Diffusion coefficients of most of foreign elements in $GaAs$ are much larger than the self-diffusion coefficients of As and Ga which can be interpreted by the movement of a pair of atoms via an arsenic vacancy.

Therefore, we must consider interstitial diffusion and other mechanisms for them. In the case of silicon in $GaAs$, silicon atoms can locate both Ga and As lattice sites, so that we must consider a diffusion mechanism based on the site transfer of silicon atoms, other than the interstitial diffusion mechanism. The site transfer diffusion was first introduced in Si-Si pair diffusion [3,19], but in the present model Si atoms need not be strongly paired.

Figure 9 illustrates the site transfer diffusion mechanism. As a first step, Si_{Ga} (silicon atom at the gallium site) transfers to V_{As} at the nearest neighbour site, resulting in the formation of Si_{As} and V_{Ga}, then V_{Ga} goes out of the lattice, or recombines with an interstitial gallium, I_{Ga}, so that thermal equilibrium is reached. As a second step, Si_{As} transfers to V_{Ga} when it comes to the nearest neighbour site, which results in the formation of Si_{Ga} and V_{As}, and the latter also returns to the thermal equilibrium. Each step can be described by the following reaction equations

(5.1) $$Si_{Ga}^+ + V_{AS}^0 + e \xrightarrow{k_1} Si_{As}^- + V_{Ga}^0 + h$$
(5.2) $$Si_{As}^- + V_{Ga}^0 + h \xrightarrow{k_3} Si_{As}^+ + V_{As}^0 + e.$$

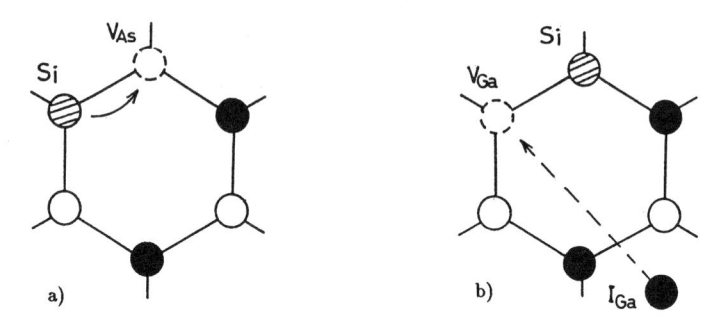

Figure 9. Illustration of the first step of the silicon migration by the site transfer mechanism.

Both reactions are equilibrated, and the equilibration can be described by the following equation.

(5.3) $$Si_{Ga}^+ + V_{As}^0 + e \underset{k_2}{\overset{k_1}{\rightleftarrows}} Si_{As}^- + V_{Ga}^0 + h; \quad \Delta G_t$$

where $\frac{k_1}{k_2} = \exp\left(-\frac{\Delta G_t}{kT}\right)$. We will later discuss the equilibrium concentration ratio $[Si_{Ga}^+]/[Si_{As}^-]$ using this equation.

Different from the case of the self-diffusion discussed in Section 3, the state corresponding to Figure 9 (b) is not a saddle point, but a stable state, because both Si_{Ga} and Si_{As} are stable. The first and the second steps occur in a series and the probabilities of occurrence are proportional to $[V_{As}]$, and $[V_{Ga}]$, respectively. That is, the rate determining step should be the latter except at very high arsenic vapor pressures because $[V_{Ga}]$ is usually much smaller than $[V_{As}]$. The formation energy of $[V_{Ga}]$ is estimated to be very high (about 3 eV as will be later discussed). Therefore, the site transfer diffusion process should be dominant at higher temperatures, while we must take into account the interstitial diffusion mechanism at lower temperatures. Let us discuss first the site transfer diffusion. Diffusion coefficient of Si_{Ga} and Si_{As} are the same for this mechanism and it can be described in the following form

(5.4) $$D(Si_{Ga}) = D(Si_{As}) = \frac{1}{6}d^2 \Gamma_{\text{eff}}$$

where Γ_{eff} is the effective jump rate, and can be described by Γ_{As} and Γ_{VGa} which are the jump rates via V_{As} and V_{Ga}, respectively, (that is, they are proportional to k_1 and k_2 in equations (5.1) and (5.2)).

(5.5) $$\Gamma_{\text{eff}}^{-1} = \Gamma_{VAs}^{-1} + \Gamma_{VGa}^{-1}$$

with

(5.6) $$\Gamma_{VAs} = \nu_{As}[V_{As}]\exp\left(-\frac{\Delta G^m_{VAs}}{kT}\right)$$

(5.7) $$\Gamma_{VGa} = \nu_{Ga}[V_{Ga}]\exp\left(-\frac{\Delta G^m_{VGa}}{kT}\right)$$

where ΔG^m_{VAs} and ΔG^m_{VGa} are corresponding free energies of migration. At low and middle vapor pressures $\Gamma_{\text{eff}} \simeq \Gamma_{VGa}$ holds, and we have,

(5.8) $$D(Si_{Ga}) = D(Si_{As}) \simeq \frac{1}{6}d^2\nu_{Ga}[V_{Ga}]\exp\left(-\frac{\Delta G^m_{VGa}}{kT}\right)$$

that is,

(5.9) $$D(Si_{Ga}) = D(Si_{As}) \simeq \frac{1}{6}d^2\nu_{Ga}(P_{As_4})^{\frac{1}{4}}\exp\left(-\frac{\Delta G^{F'}_{VGa} + \Delta G^m_{VGa}}{kT}\right).$$

As shown in Figure 10 our experiment has shown that the diffusion depth monotonically increases with the As_4 pressure at $950 \sim 1000°C$, which suggests the site transfer diffusion. However, at lower temperatures $900 \sim 875°C$, the diffusion coefficient rather decreases with increasing vapor pressure in a lower pressure region, which suggests the contribution of the interstitial diffusion mechanism as will be discussed later.

In the above expression $\Delta G^{F'}_{VGa}$ is expressed as

(2.18) $$\Delta G^{F'}_{VGa} = \Delta G^F_{VGa} - \Delta G^{\text{sub}}_{GaAs}.$$

The entalpy of the sublimation of $GaAs$ is known to be 1.14 eV. The entalpy of the formation of V_{Ga} is estimated as follows. Vacancy formation energies for silicon and diamond were estimated to be 2.3 eV and 4 eV, respectively. We assume V_{Ga} in $GaAs$ is roughly in the middle of the two values, that is, $\Delta H^F_{VGa} \simeq 3eV$ and so $\Delta H^{F'}_{VGa} \simeq 2eV$. On the other hand, the migration entalpy ΔH^m_{VGa} in Equation (5.7) is estimated as follows. In the case of silicon crystals ΔH^m_{VSi} were estimated as $\Delta H^m_{VSi} \simeq 0.33 \sim 1eV$ while, in Section 4 we have obtained for the two atom migration as $\Delta H^m \simeq 1.6eV$, so that we roughly estimated that $\Delta H^m_{VGa} \simeq 1eV$.

Therefore, our estimation of the diffusion coefficients for the site transfer mechanism is

(5.10) $$D(Si_{Ga}) = D(Si_{As}) \simeq D_0(P_{As_4})^{\frac{1}{4}}\exp\left(-\frac{3eV}{kT}\right).$$

As for the equilibrium ratio $r = [Si_{Ga}]/[Si_{As}]$, it is determined from Equation (5.3)

(5.3) $$Si_{Ga}^+ + V_{As}^0 + e = Si_{As}^- + V_{Ga}^0 + h; \quad \Delta G_t$$

as

(5.11) $$\frac{[Si_{As}]}{[Si_{Ga}^+]} \cdot \frac{p/N_v}{n/N_c} \cdot \frac{[V_{Ga}^0]}{[V_{As}^0]} = \exp\left(-\frac{\Delta G_t}{kT}\right)$$

where n and p are electron and hole concentrations, while N_c and N_v are their effective densities of states, respectively, that is, we have the relation $pn = N_c N_v \exp\left(-\frac{E_g}{kT}\right)$. In the experiment shown in Figure 10, the diffused region became n-type, that is, $[Si_{Ga}] > [Si_{As}]$ holds. If we can assume $[Si_{Ga}] \gg [Si_{As}]$, then $n \simeq N_{Ga}[Si_{Ga}^+]$ holds and we have

(5.12)
$$r = \frac{[Si_{As}^-]}{[Si_{Ga}^+]} = [Si_{Ga}^+]^2 \frac{[V_{As}^0]}{[V_{Ga}^0]} \frac{N_{Ga}^2}{N_c^2} \exp\left(-\frac{\Delta G_t - E_g}{kT}\right)$$
$$= [Si_{Ga}^+]^2 \frac{N_{Ga}^2}{N_c^2} (P_{As_4})^{-\frac{1}{4}} \exp\left(-\frac{\Delta G_t - E_g + \Delta G_{VAs}^{F'} - \Delta G_{VGa}^{F'}}{kT}\right).$$

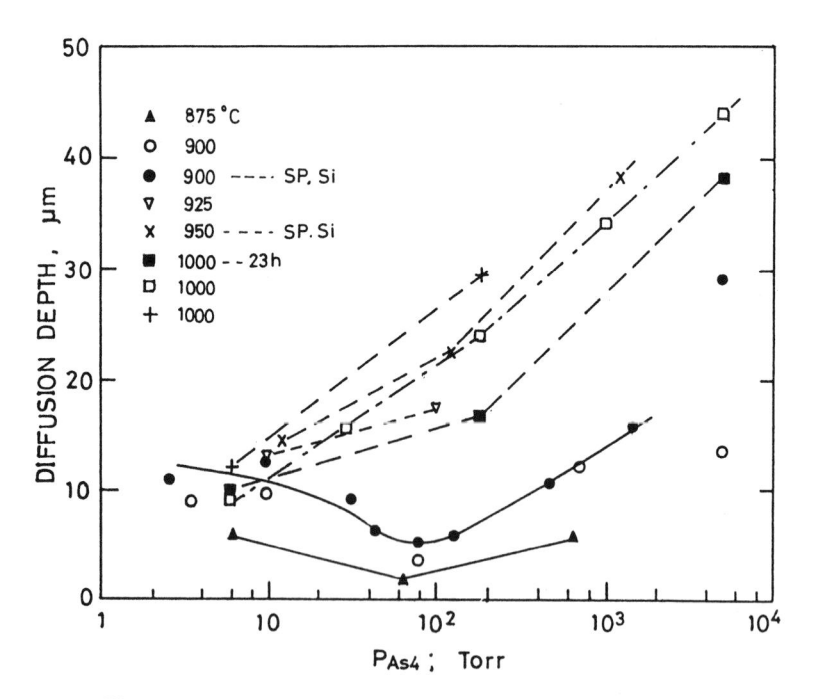

Figure 10. Experimental diffusion depth of silicon in $GaAs$ in Reference [3].

Next, we discuss the interstitial diffusion mechanism which may become dominant at lower temperatures. If we assume an isolated Si interstitial atom, I_{Si}, but not a molecular complex, the following two reaction equations will hold

(5.13)
$$Si_{Ga}^+ = I_{Si}^+ + V_{Ga}^0; \quad \Delta G_i$$
$$Si_{As}^- = I_{Si}^+ + V_{As}^0 + 2e; \quad \Delta G_i'.$$

Comparing with Equation (5.3) we have the relation

(5.14)
$$\Delta G_i - \Delta G_i' = \Delta G_t - E_g.$$

We consider the case that $[Si_{Ga}^+] \gg [Si_{As}^-]$ holds, then Equation (5.13) gives

(5.15)
$$\frac{\partial N_{\text{inter}}}{\partial N_{\text{sub}}} = \frac{[I_{Si}^+]}{[Si_{Ga}^+]} = \frac{1}{[V_{Ga}^0]} \exp\left(-\frac{\Delta G_i}{kT}\right).$$

Therefore, Equation (3.6) gives the diffusion coefficient as

(5.16)
$$D(Si_{Ga}) = D'_{\text{inter}} \frac{1}{[V_{Ga}^0]} \exp\left(-\frac{\Delta G_i}{kT}\right)$$

using Equation (5.15) and referring to Equation (3.11), we have

(5.17)
$$D(Si_{Ga}) = D'_{\text{inter}}(P_{As_4})^{-\frac{1}{4}} \exp\left(-\frac{\Delta G_i - \Delta G_{V_{Ga}}^{F'}}{kT}\right)$$
$$= \frac{1}{6}d^2\nu(P_{As_4})^{-\frac{1}{4}} \exp\left(-\frac{\Delta G_i - \Delta G_{V_{Ga}}^{F'} + \Delta G_{Si}^m}{kT}\right)$$

where ΔG_{Si}^m is the migration energy of the interstitial silicon.

On the other hand, if we consider a molecular complex $(I_{Si} \cdot I_{As})$, the corresponding reaction equation is

(5.18)
$$Si_{Ga}^+ + I_{As} = (I_{Si} \cdot I_{As})^+ + V_{Ga}; \quad \Delta G_i''$$

which gives

(5.19)
$$D(Si_{Ga}) = D'_{\text{inter}} \frac{[I_{As}]}{[V_{Ga}]} \exp\left(-\frac{\Delta G_i''}{kT}\right).$$

In this case, vapor pressure dependence is not expected because $[I_{As}]$ and $[V_{Ga}]$ have the same $(P_{As_4})^{\frac{1}{4}}$ dependence.

Experimentally, we observe the increase of the diffusion coefficient with decreasing vapor pressure at lower temperatures. Therefore we assume that silicon interstitial atoms are isolated. The entalpy parts of the free energies in Equation (5.17) are estimated as follows. As a very crude estimation, we assume that ΔH_i

is nearly equal or less than the formation entalpy of Frenkel pair in silicon crystal, that is,

$$\Delta H_i < \Delta H_{VSi}^F + \Delta H_{ISi}^F.$$

It was estimated to be about $3.1 \sim 3.3eV$ [15]. Therefore, we tentatively assume that $\Delta H_i \simeq 2.5eV$. ΔH_{Si}^m and $\Delta H_{VGa}^{F'}$ is estimated to be about $0.3eV$ and $2eV$, respectively, similar to the earlier discussions.

Then, we have the following expression for the interstitial diffusion mechanism.

$$(5.20) \qquad D(Si_{Ga}) = D_0'(P_{As_4})^{-\frac{1}{4}} \exp\left(-\frac{0.8eV}{kT}\right).$$

Figure 11 shows the calculation according to Equations (5.8) and (5.20). The constants D_0 was fitted to the experimental point at the highest vapor pressure and temperature, while D_0' was at the lowest vapor pressure and temperature.

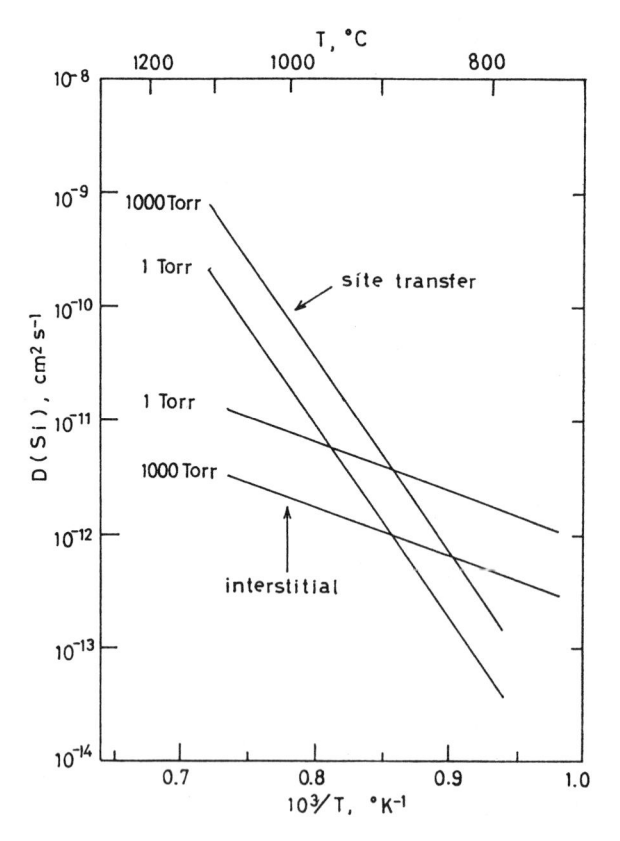

Figure 11. Calculated temperature dependence of the diffusion coefficient of silicon in $GaAs$.

As is seen in Figure 11 the both mechanism compete each other at temperatures around 900°C. As for the observed decrease in the diffusion depth around 80 Torr at 900°C, which is close to the optimum vapor pressure, it is the result of the decrease of the surface concentration of silicon atoms in the vicinity of the stoichiometric vapor pressure, rather than due to the decrease of the diffusion coefficient itself. This decrease of the surface concentration is another issue, but not discussed in this paper.

6. Diffusion at the heterostructure interface.

It was observed that the disordering of the superlattice was enhanced by the diffusion of impurities like zinc and silicon [20,21]. Let us consider the interface between $GaAs$ and $AlAs$ as illustrated in Figure 12. Different from the free surface of $GaAs$, there is no source of gallium at the heterointerface which controls the chemical potential, so that there is a sharp decrease of the gallium chemical potential in the vicinity of the heterostructure interface. This fact means that the concentrations of V_{Ga} and I_{Ga} are far from the values in equilibrium. First, we consider zinc diffusion by the interstitial mechanism.

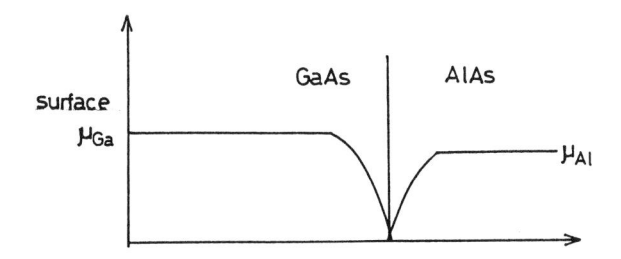

Figure 12. Schematic illustration of the changes in the gallium and aluminum chemical potentials at the $GaAs$–$AlAs$ heterostructure interface.

The reaction can be expressed as follows.

(6.1) $$Zn_s^- + 2h \rightleftarrows Zn_i^+ + V_{Ga}^0; \quad \Delta G_{Zn}.$$

In the homogeneous crystal $[V_{Ga}]$ is assumed to be equilibrated with an external arsenic vapor pressure. This equilibration is caused by the reaction equations given in Section 2, that is,

(2.16) $$GaAs \text{ (solid)} = I_{Ga} + \frac{1}{4}As_4; \quad \Delta G_{IGa}^{F'}$$

(2.17) $$\frac{1}{4}As_4 = V_{Ga} + As \text{ (lattice site)}; \quad \Delta G_{VGa}^{F'}.$$

However, it is understood that the above equation implicitly assume the existence of the free surface, which is not the case at the heterostructure interface. Actually, large amount of V_{Ga}^0 will be generated by the reaction in Equation (6.1) because ΔG_{Zn} should be much smaller than formation energies of V_{Ga} via Equation (2.17). They must recombine with I_{Ga} or go out to the free surface for the equilibration, but the concentration of I_{Ga} is much lower than the equilibrium level corresponding to the sharp decrease of the gallium chemical potential. Although a part of V_{Ga}'s go out to the free surface, but most of them contribute to the disordering at the interface.

In such a case we should consider that Equation (6.1) is equilibrated in itself, without assistance of Equations (2.16) and (2.17). That is, we have

(6.2) $$\frac{[Zn_i^+][V_{Ga}^0]}{[Zn_s^-](p/N_v)^2} = \exp\left(-\frac{\Delta G_{Zn}}{kT}\right)$$

but the condition $[Zn_i^+] = [V_{Ga}^0]$ must hold because Zn_i^+ and V_{Ga}^0 are generated as a pair.

Therefore, the concentration of V_{Ga} equilibrated by the reaction equation (6.1) is

(6.3) $$[V_{Ga}^0]^2 = [Zn_s^-](p/N_v)^2 \exp\left(-\frac{\Delta G_{Zn}}{kT}\right).$$

If we can assume that $N_{Zn_s}^- = p$, we have

(6.4) $$[V_{Ga}^0] = [Zn_s^-]^{\frac{3}{2}} \left(\frac{N_{Ga}^0}{N_v}\right) \exp\left(-\frac{\Delta G_{Zn}}{2kT}\right).$$

It is understood that $[V_{Ga}^0]$ at the interface is not determined by the arsenic vapor pressure, but determined by zinc concentration and its activation energy is $\frac{\Delta H_{Zn}}{2}$, which should be much smaller than $\Delta H_{VGa}^{F'}$, the formation entalpy in the homogeneously controlled crystal.

This excess of V_{Ga}^0 concentration is the origin of the interface disordering. In the case of silicon diffusion, both the interstitial silicon formation and the site transfer reaction can generate V_{Ga}. At lower temperatures where the interstitial diffusion is dominant, the reaction

(5.13) $$Si_{Ga}^+ = I_{Si}^+ + V_{Ga}^0; \quad \Delta G_i$$

is equilibrated. That is,

(6.5) $$\frac{[I_{Si}^+][V_{Ga}^0]}{[Si_{Ga}^+]} = \exp\left(-\frac{\Delta G_i}{kT}\right).$$

From the condition that $[I_{Si}^+] = [V_{Ga}^0]$, we have

(6.6) $$[V_{Ga}^0] = [Si_{Ga}^+]^{\frac{1}{2}} \exp\left(-\frac{\Delta G_i}{2kT}\right).$$

On the other hand, at higher temperatures where the site transfer diffusion is dominant, the reaction equation which should be equilibrated is Equation (5.3), so that we have the equilibrium relation (5.11). But this time, the condition that $[Si_{As}^-] = [V_{Ga}^0]$ must hold. As for $[V_{As}^0]$ we assume it is controlled by the external vapor pressure, that is,

$$[V_{As}^0] = (P_{As_4})^{-\frac{1}{4}} \exp\left(-\frac{\Delta G_{VAs}^{F'}}{kT}\right). \quad (2.11)$$

Therefore, we have

$$\begin{aligned}(6.7) \quad [V_{Ga}^0] &= [Si_{Ga}^+]^{\frac{2}{3}} \left(\frac{N_{Ga}^0}{N_c}\right) [V_{As}^0]^{\frac{1}{2}} \exp\left(-\frac{\Delta G_t - Eg}{2kT}\right) \\ &= [Si_{Ga}^+]^{\frac{2}{3}} \left(\frac{N_{Ga}^0}{N_c}\right) (P_{As_4})^{-\frac{1}{8}} \exp\left(-\frac{\Delta G_t - Eg + \Delta G_{VAs}^{F'}}{2kT}\right).\end{aligned}$$

Both of Equations (6.6) and (6.7) also shows that the concentration of gallium vacancies which cause the interface disordering are much increased depending on silicon concentration. We give the calculational result for Equation (6.6), which is the simplest case. As was discussed in Section 5, we assume that $\Delta H_i \simeq 2.5eV$. Also, we simply assume $\Delta S_i = 0$. As shown in Figure 13, the concentration of gallium vacancy, N_{VGa}, at the interface can be much higher than in homogeneous crystals. In the case of thermal equilibrium, N_{VGa} is expected to be of the order of $10^{11 \sim 12} cm^{-3}$ at $1000°C$, if we assume $\Delta H_{VGa} = 3eV$ and $\Delta S_{VGa} = 0$ in Equations (2.18) and (2.20). Therefore N_{VGa} shown in Figure 13 is more than 10^4 times higher than in the homogeneous crystals. These excess V_{Ga}'s in the vicinity of the interface are rapidly occupied by aluminum interstitial atoms diffusing from the $AlAs$ region driven by the steep slope of the aluminum chemical potential. The same phenomenon also occurs in the $AlAs$ region. These processes cause the gallium-aluminum mixing at the interface region.

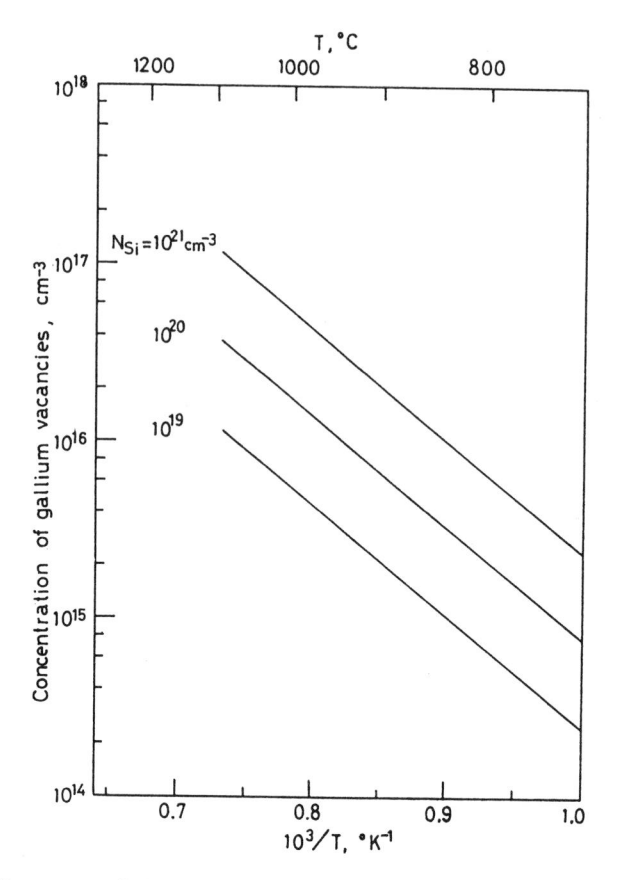

Figure 13. Calculated concentration of gallium vacancies at the heterostructure interface.

REFERENCES

[1] J. NISHIZAWA, H. OTSUKA, S. YAMAKOSHI AND K. ISHIDA, *Nonstoichiometry of Te-doped GaAs*, Jpn. J. Appl. Phys., 13 (1974), pp. 46–56.

[2] J. NISHIZAWA, I. SHIOTA AND Y. OYAMA, *Site location of As^+-ion-implanted GaAs by means of a multidirectional and high-depth-resolution Rutherford backscattering/channelling technique*, J. Phys. D: Appl. Phys., 19 (1986), pp. 1073–1078.

[3] J. NISHIZAWA, N. TOYAMA, Y. OYAMA AND K. INOKUCHI, *Influence of Arsenic Pressure on the Defects in GaAs Crystals*, Proc. of the Third Int. School on Semiconductor Optoelectronics (Cetniewo, 1981), Optoelectronic Materials and Devices ed. by M.A. Herman, PWN-Polish Scientific Publishers, Warszawa, 1983, pp. 27–77.

[4] J. NISHIZAWA, S. SHINOZAKI AND K. ISHIDA, *Properties of Sn-doped GaAs*, J. Appl. Phys., 44 (1973), pp. 1638–1645.

[5] J. NISHIZAWA, Y. OKUNO AND H. TADANO, *Nearly Perfect Crystal Growth of III-V Compounds by the Temperature Difference Method under Controlled Vapor Pressure*, J. Crystal Growth, 31 (1975), pp. 215–222.

[6] J. NISHIZAWA AND Y. OKUNO, *Stoichiometric Crystallization Method of III-V Compounds for LED's and Injection Lasers*, Proc. of Second Int. School on Semiconductor Optoelectronics (Cetniewo, 1978), Semiconductor Optoelectronics edited by M.A. Herman, PWN-Polish Scientific Publishers, Warszawa, 1980, Chap. 5, pp. 101–130.

[7] J. NISHIZAWA, Y. OKUNO AND K. SUTO, *Nearly Perfect Crystal Growth in III-V and II-VI compound semiconductors*, JARECT Vol. 19, Semiconductor Technologies (1986), edited by J. Nishizawa, OHM & North-Holland, 1986, pp. 17–80.

[8] J. NISHIZAWA, *Stoichiometry Control for Growth of III-V Crystals*, J. Crystal Growth, 99 (1990), pp. 1–8.

[9] J. NISHIZAWA, Y. OYAMA AND K. DEZAKI, *Stoichiometry-Dependent Deep Levels in n-type GaAs*, J. Appl. Phys., 67 (1990), pp. 1884–1896.

[10] J. NISHIZAWA, Y. OYAMA AND K. DEZAKI, *Formation Energy of Excess Arsenic atoms in n-type GaAs*, Phys. Rev. Letters, 65 (1990), pp. 2555–2558.

[11] ———, *Stoichiometry-Dependent Deep Levels in p-GaAs prepared by annealing under excess arsenic vapor pressure*, J. Appl. Phys., 69 (1991), pp. 1446–1453.

[12] A.B.Y. YOUNG AND G.L. PEARSON, *Diffusion of Sulfur in Gallium Phosphide and Gallium Arsenide*, J. Phys. Chem. Solids, 31 (1970), pp. 517–527.

[13] B. TUCK, *Atomic Diffusion in III-V Semiconductors*, Adam Hilger, Bristol and Philadelphia, 1988.

[14] K.K. SHIH, J.W. ALLEN AND G.L. PEARSON, *Diffusion of Zinc in Gallium Arsenide under Excess Arsenic Pressure*, J. Phys. Chem. Solids, 29 (1968), p. 379.

[15] K.H. BERNNEMANN, *New Method for Treating Lattice Points Defects in Covalent Crystals*, Phys. Rev., 137 (1965), pp. A 1497–1514.

[16] R.R. HASIGUTI, *Calculation of the Properties of Vacancies and Interstitials*, p. 27 (U.S. Government Printing Office, Washington, D.C., 1966).

[17] B. GOLDSTEIN, *Diffusion in Compound Semiconductors*, Phys. Rev., 121 (1961), pp. 1305–1311.

[18] M. KASHIWAGI, (to appear).

[19] M.E. GREINER AND J.F. GIBBONS, *Diffusion of silicon in gallium arsenide using rapid thermal processing: Experiment and model*, Appl. Phys. Lett., 44 (1984), pp. 750–752.

[20] W.D. LAIDIG, N. HOLONYAK, JR., AND M.D. CAMRAS, *Disorder of an AlAs-GaAs superlattice by impurity diffusion*, Appl. Phys. Lett., 38 (1981), pp. 776–778.

[21] K. MEEHAN, N. HOLONYAK, JR., J.M. BROWN, M.A. NIXON AND P. GAVRILOVIC, *Disorder of an $Al_xGa_{1-x}As$-GaAs superlattice by donor diffusion*, Appl. Phys. Lett. 45 (1984), pp. 549–551.

THEORY OF A STOCHASTIC ALGORITHM FOR CAPACITANCE EXTRACTION IN INTEGRATED CIRCUITS*

YANNICK L. LE COZ AND RALPH B. IVERSON[†]

Abstract. We present the theory of a novel stochastic algorithm for high-speed capacitance extraction in complex integrated circuits. The algorithm is most closely related to a statistical procedure for solving Laplace's equation known as the floating random-walk method. Our analysis begins with surface Green's functions for Laplace's equation on a scalable square domain. From them, we obtain integrals for electric potential and electric field at the domain center. An electrode-capacitance integral is next derived. This integral is expanded as an infinite sum, and probability rules that statistically evaluate the sum are deduced. These rules define the algorithm.

1. Introduction. Future technological improvements in circuit integration will make electrical connections just as important as the devices they join. It is commonly accepted that electrical performance of integrated circuits will be limited, not by device-switching speed, but by signal propagation along connection paths. Circuit designers will therefore require software "tools" that can rapidly extract capacitance, inductance, and resistance in the complex two- and three-dimensional geometries typical of integrated circuits. With these thoughts in mind, we propose a highly efficient stochastic algorithm for extracting capacitance in structures with numerous, randomly oriented electrodes.

Before discussing capacitance extraction in further detail, we will first decompose the integrated-circuit electrical connections into a set of idealized mathematical objects. For the most part the set comprises electrodes, corresponding to electrical connections themselves, and dielectrics, corresponding to various insulating layers. We must now solve Laplace's equation[1] for this system. Usually, Laplace's equation is solved for a series of electrode potentials (Dirichlet conditions), after which electric fields and inter-electrode capacitances are found. Numerical solution of Laplace's equation is generally required, since integrated-circuit geometries are complex and, to a certain extent, arbitrary. Conventional solution methods are deterministic, employing most often finite-difference, finite-element, spectral, or boundary-integral discretizations.[2-5] In two and three dimensions these methods are computationally practical as long as the geometry is relatively simple, possessing few electrodes and dielectrics. However, for complex integrated-circuit geometries these methods are computationally prohibitive.

To resolve this difficulty, we propose a random-walk algorithm that directly evaluates the inter-electrode capacitance matrix. The algorithm most closely resembles the floating random-walk method[6, 7] for solving the Laplace equation. It is particularly efficient in complex rectilinear geometries, in two and, even more so, in three dimensions. It has the added advantage of statistically estimating electric field only in regions where it is needed—the Gaussian surfaces surrounding each electrode. The fact that statistical errors in electric field tend to cancel during Gaussian-surface integration enhances algorithm efficiency as well. Importantly, when evaluating the capacitance matrix, our algorithm requires a number of Gaussian-surface integra-

---

* Written material in this paper has been excerpted from a larger work in draft form which will be submitted to *Solid State Electronics*, for future publication

[†] Department of Electrical, Computer, and Systems Engineering, Rensselaer Polytechnic Institute, Troy, NY 12180-3590

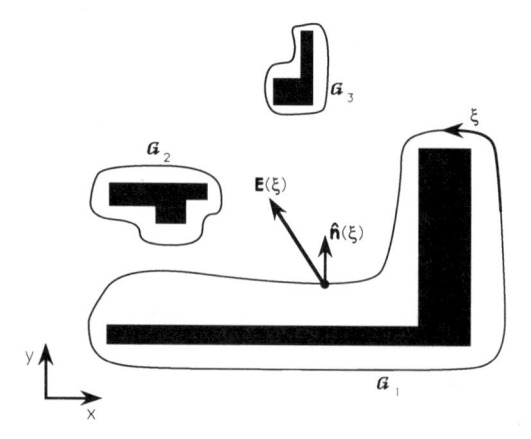

FIG. 1. *An example of a two-dimensional, rectilinear electrode arrangement. Gaussian boundaries are denoted $\mathcal{G}_1, \mathcal{G}_2$, and $\mathcal{G}_3$. The parameter ξ measures length along any of these boundaries. Note also, electric-field $\mathbf{E}$ and outward unit normal $\hat{n}$ both depend on ξ.*

tions equal to the total electrode number. Usual capacitance-extraction procedures, in contrast, require on order of the square of such as many integrations. We note lastly that, owing to its stochastic nature, the algorithm readily parallelizes for speed improvement.

Hereafter, we will assume a two-dimensional rectilinear geometry and neglect variation in electric permittivity. The sections that follow constitute the theory of our algorithm for capacitance extraction. We will first derive a nested-integral expansion for the capacitance matrix associated with a general assembly of rectilinear electrodes. To efficiently evaluate this expansion, we then deduce a novel stochastic algorithm. We finish with a proof of the algorithm's mathematical validity.

2. Capacitance-integral expansion. We will now deduce an expression for the electrode-capacitance matrix. Figure 1 is an example electrode arrangement. These two-dimensional electrodes are shown in black, their edges parallel to the xy-coordinate axes. It is understood that the electrodes extend infinitely in the $+z$ and $-z$ directions. For N electrodes, off-diagonal elements of the capacitance-matrix C_{ij}, $i \neq j$,[1] are defined according to

$$(1) \qquad q_i = \sum_{j=1}^{N} C_{ij}(v_i - v_j),$$

where $i = 1, \ldots, N$. Above, the $v_1, \ldots, v_N$ and $q_1, \ldots, q_N$ denote electrode voltages and their corresponding charges per unit length in z. Appropriately, C_{ij} is a capacitance per unit length in z as well.

Gauss's law permits us to write

$$(2) \qquad q_i = \epsilon \int_{\mathcal{G}_i} d\zeta \; \mathbf{E}(\xi) \cdot \hat{n}(\xi),$$

---

[1] Diagonal elements $C_{11}, C_{22}, \ldots$ represent, what we call, electrode "self-capacitances". They serve no use in electrical modeling.

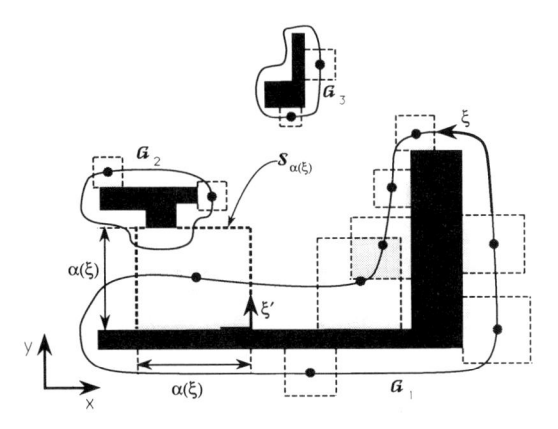

FIG. 2. *Examples of initial maximal squares. Boundaries $S_{\alpha(\xi)}$ are of edge size $\alpha(\xi)$ and are centered on Gaussian-boundary points (dark dots).*

where ϵ is a constant electric permittivity and $\mathcal{G}_i$ is a Gaussian boundary (actually a Gaussian-surface cross section) surrounding a single electrode i. The parameter ξ measures boundary length on which electric field $\boldsymbol{E}$ and outward unit normal $\hat{\boldsymbol{n}}$ depend. Figure 1 shows this parameterization.

We can rewrite (2) by replacing each component of $\boldsymbol{E}$ with its integral equivalent. We will express these components as an integral along the edges of a square $\mathcal{S}_{\alpha(\xi)}$, parallel to the coordinate axes and centered about any particular boundary point on $\mathcal{G}_i$. For each ξ, the square will be chosen the largest possible containing *no* electrodes. The square's boundaries thus conform, at least in part, to the electrode boundaries themselves. Figure 2 gives examples of such constructions.

Denoting the edge size of these maximal squares as $\alpha(\xi)$, we expand (2) as a double integral. To this end,

$$(3) \qquad q_i = -\int_{\mathcal{G}_i} d\xi \; s_i(\xi) \int_{\mathcal{S}_{\alpha(\xi)}} d\xi' \; w(\xi|\xi') \, \bar{G}[\alpha(\xi)|\xi')] \, \psi_{\mathcal{S}_{\alpha(\xi)}}(\xi'),$$

where at each electrode $i = 1, \cdots, N$ we have defined a weight function

$$(4) \qquad w(\xi|\xi') = \epsilon \, \frac{\hat{n}_x(\xi) \, \bar{G}_x[\alpha(\xi)|\xi'] + \hat{n}_y(\xi) \, \bar{G}_y[\alpha(\xi)|\xi']}{s_i(\xi) \bar{G}[\alpha(\xi)|\xi']}.$$

Above, $\hat{n}_x$ and $\hat{n}_y$ are components of $\hat{\boldsymbol{n}}$. The quantities $\bar{G}, \bar{G}_x$, and $\bar{G}_y$ are, respectively, suitably defined surface Green's functions[1] for electric potential, x component of electric field, and y component of electric field. We have also introduced, for completeness, arbitrary sampling-functions $s_i(\xi)$, which do not affect the value of charge-integrals (3). These functions are assumed normalized over the $\mathcal{G}_i$,

$$(5) \qquad \int_{\mathcal{G}_i} d\xi \; s_i(\xi) = 1.$$

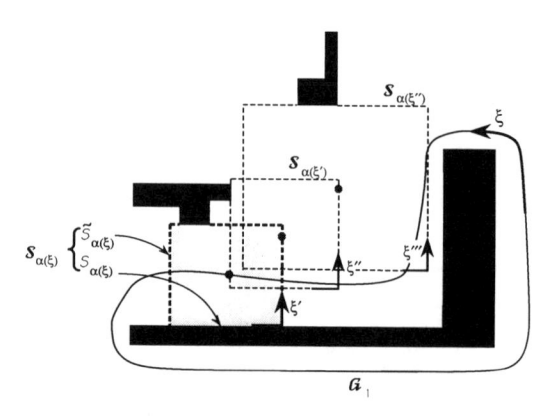

FIG. 3. *Examples of subsequent maximal squares (initial square shaded). Each boundary $S_{\alpha(\xi^{\cdots\prime})}$ can be decomposed into an electrode part $S_{\alpha(\xi^{\cdots\prime})}$ and a non-electrode part $\tilde{S}_{\alpha(\xi^{\cdots\prime})}$. These squares are centered on previous square-boundary points (dark dots).*

To deduce our capacitance algorithm, we must express the q_i in terms of electrode potentials. We begin by splitting the domain $S_{\alpha(\xi)}$ into two parts, examples of which are shown in Fig. 3. The first $S_{\alpha(\xi)}$ is the electrode part, and the second $\tilde{S}_{\alpha(\xi)}$ is the nonelectrode part. Equation (3) can then be written as a sum of two integrals, one over $S_{\alpha(\xi)}$ and the other over $\tilde{S}_{\alpha(\xi)}$. The nonelectrode potential $\psi_{\tilde{S}_{\alpha(\xi)}}(\xi')$ in the latter integral can be replaced. We do so with the aid of $\bar{G}$. Note carefully, we must use a new dummy variable ξ'', and we choose[2] $S_{\alpha(\xi')}$ as the new integration domain. Also, we have extended our domain-construction procedure as shown in Fig. 3. Geometrically, $S_{\alpha(\xi')}$ is the largest square, containing no electrodes, centered about any particular boundary point on $\tilde{S}_{\alpha(\xi)}$. If this entire process—splitting $S_{\alpha(\xi')}$ into $S_{\alpha(\xi')}$ and $\tilde{S}_{\alpha(\xi')}$, expanding as a sum of two integrals, and replacing the non-electrode-boundary potential by means of $\bar{G}$—is repeated indefinitely, one obtains an infinite sum of nested integrals:[3]

---

[2] To simplify notation, we supress the dependence of $\alpha(\xi')$ on ξ. In general, we have $\alpha(\xi^{\cdots\prime}) = \alpha(\xi, \xi', \ldots, \xi^{\cdots\prime})$.

[3] Higher-order additive terms in (6) are generated with a simple recursive sequence: (i) copy the last known term, (ii) place a tilde over its rightmost integral limit, (iii) replace $\psi_{S_{\alpha(\xi^{\cdots\prime})}}(\xi^{\cdots\prime\prime})$ within it by $\int_{S_{\alpha(\xi^{\cdots\prime\prime})}} d\xi^{\cdots\prime\prime\prime}\, \bar{G}[\alpha(\xi^{\cdots\prime\prime})|\xi^{\cdots\prime\prime\prime}] \psi_{S_{\alpha(\xi^{\cdots\prime\prime})}}(\xi^{\cdots\prime\prime\prime})$.

$$q_i = -\int_{\mathcal{G}_i} d\xi\, s_i(\xi) \left\{ \vphantom{\int_{\tilde{S}_{\alpha(\xi)}}} \right.$$

$$\int_{S_{\alpha(\xi)}} d\xi'\, w(\xi|\xi')\, \bar{G}[\alpha(\xi)|\xi']\psi_{S_{\alpha(\xi)}}(\xi')$$

$$+ \int_{\tilde{S}_{\alpha(\xi)}} d\xi'\, w(\xi|\xi')\, \bar{G}[\alpha(\xi)|\xi'] \int_{S_{\alpha(\xi')}} d\xi''\, \bar{G}[\alpha(\xi')|\xi'']\psi_{S_{\alpha(\xi')}}(\xi'')$$

$$+ \int_{\tilde{S}_{\alpha(\xi)}} d\xi'\, w(\xi|\xi')\, \bar{G}[\alpha(\xi)|\xi'] \int_{\tilde{S}_{\alpha(\xi')}} d\xi''\, \bar{G}[\alpha(\xi')|\xi''] \times$$

$$\int_{S_{\alpha(\xi'')}} d\xi'''\, \bar{G}[\alpha(\xi'')|\xi''']\psi_{S_{\alpha(\xi'')}}(\xi''')$$

(6) $$+ \quad \cdots \quad \left. \vphantom{\int} \right\}.$$

Expansion (6) depends only on electrode potentials. Grounding all electrodes except the jth, which we set to some arbitrary voltage v_j, reduces (1) to

(7) $$C_{ij} = -\frac{q_i}{v_j},$$

for $i = 1,\ldots,N$ ($i \neq j$). Since the q_i of (6) are linear functions of v_j, C_{ij} can be written as an integral expansion independent of v_j. This argument is valid for any possible j; thus, off-diagonal elements of the capacitance matrix are independent of electrode voltages, and we have

$$C_{ij} = \int_{\mathcal{G}_i} d\xi\, s_i(\xi) \left\{ \vphantom{\int} \right.$$

$$\int_{S^j_{\alpha(\xi)}} d\xi'\, w(\xi|\xi')\, \bar{G}[\alpha(\xi)|\xi']$$

$$+ \int_{\tilde{S}_{\alpha(\xi)}} d\xi'\, w(\xi|\xi')\, \bar{G}[\alpha(\xi)|\xi'] \int_{S^j_{\alpha(\xi')}} d\xi''\, \bar{G}[\alpha(\xi')|\xi'']$$

$$+ \int_{\tilde{S}_{\alpha(\xi)}} d\xi'\, w(\xi|\xi')\, \bar{G}[\alpha(\xi)|\xi'] \int_{\tilde{S}_{\alpha(\xi')}} d\xi''\, \bar{G}[\alpha(\xi')|\xi''] \int_{S^j_{\alpha(\xi'')}} d\xi'''\, \bar{G}[\alpha(\xi'')|\xi''']$$

(8) $$+ \quad \cdots \quad \left. \vphantom{\int} \right\}.$$

The boundary[4] $S^j_{\alpha(\xi^{\cdots\prime})}$ is simply the portion of $S_{\alpha(\xi^{\cdots\prime})}$ coincident with electrode j.

3. Extraction algorithm and proof. In practice, direct evaluation of (8) is computationally prohibitive, especially for integrated-circuit geometries where N is usually large. We resolve this difficulty with a stochastic algorithm that estimates the C_{ij}:

1. Partition each integration variable in (8) into small segments of, possibly different, size $\Delta\xi_r^{\cdots\prime}$ ($r = 1,2,\ldots$). Define corresponding discrete variables at each segment center $\xi_r^{\cdots\prime}$.

---

[4] We mean here the superscript '$\cdots\prime$' to possibly include the unprimed case. That is, $\xi^{\cdots\prime}$ is one of $\xi, \xi', \xi'', \ldots$.

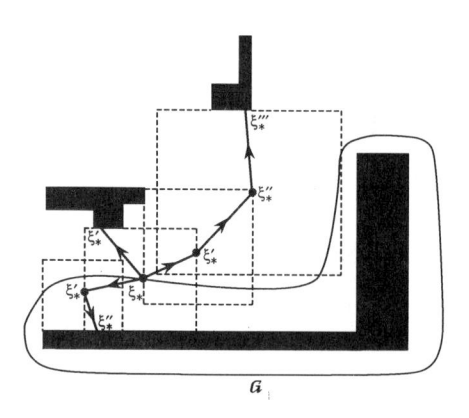

FIG. 4. *Examples of first-, second-, and third-order walks. For clarity, the walks have been drawn to start at the same boundary-point ξ_* of $\mathcal{G}_1$.*

2. Introduce new variables $\mathcal{N}_i$ and $\mathcal{W}_i$, the meaning of which will be made clear shortly. Initially, set $\mathcal{N}_i = \mathcal{W}_{ij} = 0$ for all ij $(i,j = 1,\ldots,N)$. Set $i = 1$.
3. Randomly pick a ξ_r, say ξ_*, on $\mathcal{G}_i$, with discrete probability distribution $P_\xi(\xi_r) = \int_{\Delta\xi_r} d\xi\ s_i(\xi)$.
4. Randomly pick a ξ'_r, say ξ'_*, on $\mathcal{S}_{\alpha(\xi_*)}$, with discrete probability distribution, conditioned by ξ_*, $P_{\xi'}(\xi_*|\xi'_r) = \int_{\Delta\xi'_r} d\xi'\ \bar{G}[\alpha(\xi_*)|\xi']$.
5. If the last variable picked in Step 4 is not on an electrode boundary, then change Step 4 as follows: mark *every* occurrence of ξ with an additional prime[5] ' ' ' and repeat Step 4.
6. If the last variable picked in Step 4 is on the jth electrode boundary, then replace $\mathcal{N}_i$ with $\mathcal{N}_i + 1$ and $\mathcal{W}_{ij}$ with $\mathcal{W}_{ij} + w(\xi_*|\xi'_*)$.
7. If $\mathcal{N}_i$ is sufficiently large, go to Step 8. Else, change Step 4 to its original form (written here above) and go to Step 3.
8. $C_{ij} = \mathcal{W}_{ij}/\mathcal{N}_i$ for $j = 1,\ldots,N$. If $i = N$, then stop. Else, replace i with $i+1$, change Step 4 to its original form (written here above), and go to Step 3.

We will now explain why this capacitance-extraction algorithm works. For a given starting electrode i, enumerate a possible set of $\mathcal{N}_i$ random trajectories, or "walks", generated by the algorithm that start on $\mathcal{G}_i$ and end on any electrode. Figure 4 gives examples of such first-order ($\xi_* \to \xi'_*$), second-order ($\xi_* \to \xi'_* \to \xi''_*$), and third-order ($\xi_* \to \xi'_* \to \xi''_* \to \xi'''_*$) walks.

---

[5] This applies to *all* occurences of ξ, regardless of subscripts and superscripts. For example, $P_{\xi'}(\xi_*|\xi'_r) = \int_{\Delta\xi'_r} d\xi'\ \bar{G}[\alpha(\xi_*|\xi')] \to P_{\xi''}(\xi'_*|\xi''_r) = \int_{\Delta\xi''_r} d\xi''\ \bar{G}[\alpha(\xi'_*|\xi'')]$. Once Step 4 has been properly changed it *remains* so until otherwise stated.

Note also, in keeping with footnote 2, $\mathcal{S}_{\alpha(\xi_*^{\cdots\prime})} = \mathcal{S}_{\alpha(\xi_*,\xi'_*,\ldots,\xi_*^{\cdots\prime})}$ in Step 4, where $\xi_*, \xi'_*, \ldots, \xi_*^{\cdots\prime}$ are the set of random picks *before* entering or re-entering Step 4.

Consider, for the moment, a subset of this enumeration consisting of first-order walks ending on a specific electrode j. Of the $\mathcal{N}_i$ total walks starting from $\mathcal{G}_i$, $\mathcal{N}_i P_\xi(\xi_r)$ of them start at ξ_r. Of those, a fraction $P_{\xi'}(\xi_r|\xi_r')$ end at ξ_r' on $S^j_{\alpha(\xi_r)}$. Therefore, the total number of walks from ξ_r to ξ_r' is simply $\mathcal{N}_i P_\xi(\xi_r) P_{\xi'}(\xi_r|\xi_r')$. The algorithm sums $w(\xi_r|\xi_r')$ over all possible ξ_r,ξ_r'-pairs and divides by $\mathcal{N}_i$, giving

$$(9) \quad C_{ij}^{(1)} \approx \frac{\mathcal{W}_{ij}^{(1)}}{\mathcal{N}_i} = \sum_{\mathcal{G}_i} P_\xi(\xi_r) \sum_{S^j_{\alpha(\xi_r)}} w(\xi_r|\xi_r') P_{\xi'}(\xi_r|\xi_r').$$

The sums in (9) are to be taken over discrete-points ξ_r and ξ_r' on their respective surfaces $\mathcal{G}_i$ and $S^j_{\alpha(\xi_r)}$. In addition, we have designated first-order-walk contributions with the superscript '(1)'. Observe that (9) and the discussion preceding it are valid for any starting Gaussian surface and ending electrode, that is, any ij-pair ($i \neq j$). Remember, as well, that (9) is a good approximation to $C_{ij}^{(1)}$ when $\mathcal{N}_i$ is large—large enough to ensure that the known probability-distributions P_ξ and $P_{\xi'}$ adequately represent the *actual* distributions in our enumeration of walks.

A connection with (8) follows upon replacing P_ξ and $P_{\xi'}$ in (9) with their integral equivalents from Steps 3 and 4 of the algorithm. We get

$$(10) \quad C_{ij}^{(1)} \approx \sum_{\mathcal{G}_i} \left[\int_{\Delta \xi_r} d\xi\; s_i(\xi) \right] \sum_{S^j_{\alpha(\xi_r)}} w(\xi_r|\xi_r') \left\{ \int_{\Delta \xi_r'} d\xi'\; \bar{G}[\alpha(\xi_r)|\xi'] \right\}.$$

The $\Delta \xi_r'$ are assumed small enough so that $w(\xi_r|\xi_r')$ varies little over their extent. This permits us to change ξ_r' to ξ' in w and to move w within the rightmost integrand of (10). Hence,

$$(11) \quad C_{ij}^{(1)} \approx \sum_{\mathcal{G}_i} \left[\int_{\Delta \xi_r} d\xi\; s_i(\xi) \right] \sum_{S^j_{\alpha(\xi_r)}} \int_{\Delta \xi_r'} d\xi'\; w(\xi_r|\xi') \bar{G}[\alpha(\xi_r)|\xi'].$$

Now, evaluating the rightmost sum above, we find immediately

$$(12) \quad C_{ij}^{(1)} \approx \sum_{\mathcal{G}_i} \left[\int_{\Delta \xi_r} d\xi\; s_i(\xi) \right] \int_{S^j_{\alpha(\xi_r)}} d\xi'\; w(\xi_r|\xi') \bar{G}[\alpha(\xi_r)|\xi'].$$

Lastly, if we assume the $\Delta \xi_r$ are small enough, so that the rightmost integral in (12) varies little over their extent, we can change ξ_r to ξ in $w, \bar{G}$, and α; and move the integral within the leftmost integrand. Evaluating the remaining sum, as before, gives our final result:

$$(13) \quad C_{ij}^{(1)} \approx \int_{\mathcal{G}_i} d\xi\; s_i(\xi) \int_{S^j_{\alpha(\xi)}} d\xi\; w(\xi|\xi') \bar{G}[\alpha(\xi)|\xi'].$$

The expression above is a first-order approximation to C_{ij}, in other words, the first term in expansion (8). The general proof for the nth term, $C_{ij}^{(n)}$, in (8) proceeds alongs lines similar to (9)–(13) and is left to the reader.

The algorithm actually sums w for walks of varying order n. Consequently, it generates a statistical estimate of C_{ij} by summing statistical estimates of $C_{ij}^{(n)}$ over n. Mathematically, we have

$$(14) \quad C_{ij} \approx \sum_{n=1}^\infty C_{ij}^{(n)} = \frac{1}{\mathcal{N}_i} \sum_{n=1}^\infty \mathcal{W}_{ij}^{(n)} = \frac{\mathcal{W}_{ij}}{\mathcal{N}_i},$$

where $\mathcal{W}_{ij}^{(n)}$ is the sum of weight-functions w for all nth-order random walks starting on Gaussian boundary $\mathcal{G}_i$ and ending on electrode j.

Acknowledgement. The authors thank Professor James D. Meindl for sponsoring this work and for helpful technical discussions. The authors also thank Professor Alan L. McWhorter for reading the original manuscript and suggesting improvements.

REFERENCES

[1] P.M. Morse and H. Feshbach, *Methods of Theoretical Physics*, Part I, McGraw-Hill, New York, 1953.

[2] A.H. Zemanian, "A Finite-Difference Procedure for the Exterior Problem Inherent in Capacitance Computations for VLSI Interconnections", *IEEE Trans. Electron Devices*, vol. 35, pp. 985–991, 1988.

[3] P.E. Cottrell and E.M. Buturla, "VLSI Wiring Capacitance", *IBM J. Res. Develop.*, vol. 29, pp. 277–288, 1985.

[4] F.S. Lai, "Coupling Capacitances in VLSI Circuits Calculated by Multi-Dimensional Discrete Fourier Series", vol. 32, pp. 141–148, 1989.

[5] A.E. Ruehli and P.A. Brennan, "Efficient Capacitance Calculations for Three-Dimensional Multiconductor Systems", *IEEE Trans. Microwave Theory Tech.*, vol. MTT-21, pp. 76–82, 1973.

[6] G.M. Brown, "Monte Carlo Methods" in *Modern Mathematics for Engineers*, E.F. Beckenbach, editor, McGraw-Hill, New York, 1956.

[7] A. Haji-Sheikh and E.M. Sparrow, "The Solution of Heat Conduction Problems by Probability Methods", *Trans. ASME*, vol. C-89, pp. 121–131, 1967. (See, in particular, the section "Authors Closure", and references therein.)

MOMENT-MATCHING APPROXIMATIONS FOR LINEAR(IZED) CIRCUIT ANALYSIS[*]

NANDA GOPAL, ASHOK BALIVADA AND LAWRENCE T. PILLAGE[†]

Abstract. Moment-matching approximations appear to be a promising approach for linear circuit analysis in several application areas. Asymptotic Waveform Evaluation (AWE) uses moment-matching to approximate the time- or frequency-domain circuit response in terms of a reduced-order model. AWE has been demonstrated as an efficient means for solving large, stiff, linear(ized) circuits, in particular, the large RC- and RLC-circuit models which characterize high-speed VLSI interconnect. However, since it is based upon moment-matching, which has been shown to be equivalent to a Padé approximation in some cases, AWE is prone to yielding unstable waveform approximations for stable circuits. In addition, it is difficult to quantify the time domain error for moment-matching approximations. We address the issues of stability and accuracy of moment-matching approximations as they apply to linear circuit analysis.

1. Introduction. Moment-matching approximations appear to be a promising approach for linear circuit analysis in several application areas. Asymptotic Waveform Evaluation (AWE) [20] uses moment-matching to approximate the time- or frequency-domain response of an n-th order circuit in terms of a reduced qth order model. For large, linear RLC circuits with thousands of poles, the response at any node tends to be *dominated* by only a few of the poles, therefore excellent approximations are possible for $q << n$. The *dominant poles* also tend to be the lower frequency poles (poles closer to the origin) which contain most of the signal energy. The effects due to higher frequency poles that influence the response only for a very short time and contain little energy [7] are *averaged* into the waveform approximation. The order of the AWE approximation – the number of dominant pole(s) in the approximation – determines the overall waveform accuracy.

AWE has been demonstrated as an efficient means for solving, large, stiff, linear circuits, in particular, the large RC- and RLC-interconnect circuit models that are difficult to evaluate using traditional circuit simulation algorithms. However, since it is based upon moment-matching, which has been shown to be equivalent to a Padé approximation in some cases, AWE is prone to yielding unstable waveform approximations for stable circuits. In addition, it is difficult to quantify the time-domain error which makes it difficult to select an appropriate order for the moment-matching approximation. In this paper we consider these stability and accuracy issues as they apply to linear circuit analysis and we propose some techniques for addressing these problems for linear, passive RLC circuits.

2. Moment-matching methods. The use of moments in the simplification of high-order systems is a well-known technique [3, 9, 26]. Moment values of actual, physical, linear-systems can be obtained from experimental data, calculated from an exact transfer function expression, or measured from a model of the system. From 2q moment values a q-th order model can be uniquely specified.

---

[*] This work was supported by the National Science Foundation under the grant MIP #9007917.

[†] Computer Engineering Research Center, Department of Electrical & Computer Engineering, The University of Texas at Austin, Austin, Texas 78712

Model-order reduction via moment-matching is motivated by the characteristics we seek in a good system model approximation. For an asymptotically stable system with a proper, rational, transfer function $H(s)$, or impulse response $h(t)$, Zakian [26] defines a "good" system model, $\hat{H}(s)$ or $\hat{h}(t)$, as one for which:

1. $[h(t) - \hat{h}(t)]$ converges rapidly to 0 as $t \to \infty$
2. $h(0) - \hat{h}(0) = 0$
3. $\max_{(0 < t \leq \infty)} \left| h(t) - \hat{h}(t) \right| \leq K > 0$

where K is a constant which is application dependent.

Condition 2 is easily satisfied by forcing the Initial Value Theorem to apply. To ensure Condition 1, one potential criterion is:

$$\text{(1)} \qquad \int_0^\infty t^j [h(t) - \hat{h}(t)] dt = 0, \; j = 0, 1, 2, \ldots, (m+n)$$

where m and n are respectively the degrees of the numerator and denominator of $\hat{H}(s)$. Equation (1) is recognized to suggest that the first $(m+n+1)$ moments of h and $\hat{h}$ are equal. Finally, it can be shown that Condition 3 is satisfied for sufficiently large values of $(m+n)$ [26].

The existence of the moments of the actual system function $h(t)$ is ensured if $h(t)$ is piecewise continuous in $[0, \infty)$ and of exponential order $O[exp(\sigma t)], t \to \infty, \sigma < 0$ [26]. For passive, linear RLC circuits, which are asymptotically stable, the responses are smooth and piecewise continuous in $[0, \infty)$, thus satisfying both the above requirements.

The preceding discussion developed a reduced-order model $\hat{H}(s)$, that is termed a moment-approximant. The Padé approximants are a similar class of approximating functions that are related to moment-approximants. A Padé approximant, denoted $[P/Q]$, is a rational function approximation of a transfer function $H(s)$, analytic about $s = 0$, such that the first $(P+Q+1)$ coefficients of the MacLaurin expansions of $[P/Q]$ and $H(s)$ are equal [1, 2]. In the above definition, P and Q refer to the degrees of the numerator and denominator polynomials respectively, in the Padé approximant.

To establish the relation between the moment- and the Padé-approximants, consider the Laplace transform definition of an analytic function $h(t)$:

$$\text{(4)} \qquad H(s) = \int_0^\infty e^{-st} h(t) dt$$

If, for any function $f(t)$, the integral

$$\text{(2)} \qquad \int_0^\infty f(t) dt$$

exists, the nth moment M_n of the function about the origin is defined as [9]

$$\text{(3)} \qquad M_n = \int_0^\infty t^n f(t) dt$$

It is shown in [9, 17] that the normalized moments M_i/M_0 are analogous to the mean of a probability distribution function.

Expanding e^{-st} in a MacLaurin series yields:

$$H(s) = \sum_{j=0}^{\infty} \frac{1}{j!}(-s)^j \int_0^{\infty} t^j h(t) dt \tag{5}$$

$$= \sum_{j=0}^{\infty} \frac{(-1)^j}{j!} M_j s^j \tag{6}$$

In other words, the time moments of a function $h(t)$ are related to the coefficients of the MacLaurin series expansion of $h(t)$. The following theorem by Zakian [26], explicitly defines the relation between the moment- and the Padé-approximants:

THEOREM 2.1. *Let h be piecewise continuous on $[0, \infty)$ and of exponential order $O[exp(\sigma t)]$, $t \to \infty$, $\sigma < 0$, and let the Laplace transform $\mathcal{L}\{\hat{h}\}$ be an asymptotically stable (m/n) rational function; then $\hat{h}$ is the $(m+n)$ moment-approximant of h if and only if $\mathcal{L}\{\hat{h}\}$ is the Padé approximant $[m/n]$ of $\mathcal{L}\{h\}$.* The reference to $\mathcal{L}\{\hat{h}\}$ being asymptotically stable is an important one, and will be addressed in more detail in a later section on stability.

3. Asymptotic waveform evaluation. Asymptotic Waveform Evaluation (AWE) is a generalized approach to approximating the waveform response of linear(ized) RLC circuits via moment-matching [19, 20]. AWE is most conveniently explained in terms of the differential state equations for a lumped, linear, time-invariant (LLTI) circuit:

$$\dot{\vec{x}} = \mathbf{A}\vec{x} + \vec{b}\vec{\delta} \tag{7}$$

where $\vec{x}$ is the n-dimensional state vector and $\vec{\delta}$ is an m-dimensional excitation vector of impulses. Such a circuit description can be found for most LLTI circuits. Modeling the state variables permits the modeling of any output variable as a linear combination of these state variables. While the following development can be applied to other excitation forms such as step or ramp voltages, we consider only impulse excitations since from them, all other responses can be obtained by analytical convolution and superposition.

The Laplace transform solution of Eq.(7) is

$$\vec{X}(s) = (s\mathbf{I} - \mathbf{A})^{-1}\vec{b} \tag{8}$$

which can be expanded into a MacLaurin series

$$\vec{X}(s) = -\mathbf{A}^{-1}(\mathbf{I} + \mathbf{A}^{-1}s + \mathbf{A}^{-2}s^2 + \ldots)\vec{b} \tag{9}$$

Focusing on a specific component of $\vec{X}(s)$, say the ith, the coefficients of the series expansion can be denoted as:

$$\begin{aligned} m_0^i &= [-\mathbf{A}^{-1}\vec{b}]_i \\ m_1^i &= [-\mathbf{A}^{-2}\vec{b}]_i \\ &\vdots \\ m_{2q-1}^i &= [-\mathbf{A}^{-2q}\vec{b}]_i \end{aligned} \tag{10}$$

where m_j^i denotes the j-th coefficient in the series expansion of the i-th state variable.

The efficiency of AWE lies in the recursive computation of these coefficients $[\vec{m}]_i$. As explained in [20], the explicit construction/inversion of the state matrix $\mathbf{A}$ is not required. Instead, finding $\mathbf{A}^{-1}$ from Eq.(7) is equivalent to solving for the port voltages of the open-circuit capacitance ports and port currents of the short-circuit inductance ports [6]. To illustrate, consider the circuit in Fig.1(a). The response

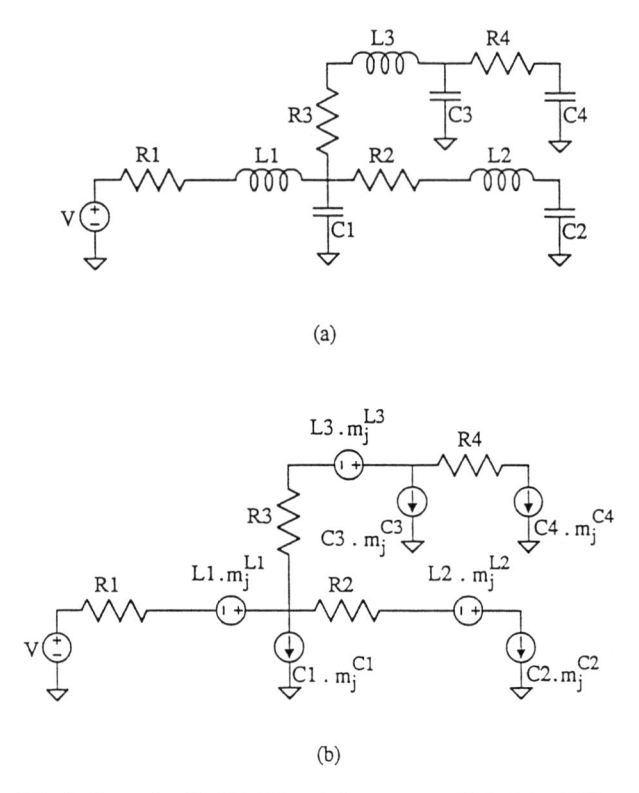

FIG. 1. *Computing the $(j+1)$th set of response coefficients in AWE.*

coefficients can be obtained from the circuit in Fig.1(b), where all the capacitors have been replaced by current sources, and inductors by voltage sources.

The recursion in Eq.(10) is initiated by replacing the source V in Fig.1(b), by a constant voltage of value 1, and setting all the capacitor and inductor-sources to zero. Solving this dc circuit for the capacitor voltages and inductor currents yields the first coefficient m_0 for each state variable. This is equivalent to substituting $\vec{\delta} = 1$, $\dot{\vec{x}} = 0$ in Eq.(7), and solving for $\vec{x}$. This yields $\vec{x} = -\mathbf{A}^{-1}\vec{b}$, which, from Eq.(10), is the first coefficient m_0.

Higher-order coefficients m_j are then recursively obtained by shorting the excitation source V, setting the capacitor-current sources equal to $-C_i m^i_{j-1}$, the inductor-voltage sources equal to $-L_i m^i_{j-1}$, and solving for the port voltages and currents. This is again equivalent to setting $\vec{\delta} = 0$ in Eq.(7), $\dot{\vec{x}} = \vec{x}_{j-1}$ and solving for $\vec{x}_j$ [13, 20].

In AWE, the reduced qth-order model of the ith state variable has the form:

$$[\hat{X}(s)]_i = \sum_{l=1}^{q} \frac{k_l^i}{(s - p_l^i)} \tag{11}$$

where the terms p_l^i are the q, unique dominant-pole approximations and the terms k_l^i are the corresponding residues. The values of p_l^i and k_l^i are computed such that the model in Eq.(11) best approximates the actual response in Eq.(9) in the sense of the Padé approximation:

$$
\begin{aligned}
m_0^i + m_1^i s + m_2^i s^2 + \ldots &= \sum_{l=1}^{q} \frac{k_l^i}{(s - p_l^i)} + O(s^{2q}) \\
&\approx \frac{b_0 + b_1 s + \cdots + b_{q-1} s^{q-1}}{1 + a_1 s + \cdots + a_{q-1} s^{q-1} + a_q s^q}
\end{aligned}
\tag{12}
$$

Cross-multiplying and equating the coefficients of s^q, s^{q+1}, ..., yields the following set of linear equations for the denominator coefficients of Eq.(12) [13, 20]:

$$
\begin{bmatrix}
m_0^i & m_1^i & \cdots & m_{q-1}^i \\
m_1^i & m_2^i & \cdots & m_q^i \\
\vdots & \vdots & \ddots & \vdots \\
m_{q-1}^i & m_q^i & \cdots & m_{2q-2}^i
\end{bmatrix}
\begin{bmatrix}
a_q \\ a_{q-1} \\ \vdots \\ a_1
\end{bmatrix}
=
\begin{bmatrix}
m_q^i \\ m_{q+1}^i \\ \vdots \\ m_{2q-1}^i
\end{bmatrix}
\tag{13}
$$

The roots of the characteristic polynomial

$$1 + a_1 s + a_2 s^2 + \cdots + a_{q-1} s^{q-1} + s^q = 0 \tag{14}$$

are the dominant pole approximations.

To solve for the corresponding residues $\vec{k}^i$, the first q coefficients of the s terms in the expansion of Eq.(11) are matched to those of Eq.(9) to obtain the system:

$$
\begin{aligned}
-\left(\frac{k_1^i}{p_1^i} + \frac{k_2^i}{p_2^i} + \ldots + \frac{k_q^i}{p_q^i}\right) &= m_0^i \\
-\left(\frac{k_1^i}{(p_1^i)^2} + \frac{k_2^i}{(p_2^i)^2} + \ldots + \frac{k_q^i}{(p_q^i)^2}\right) &= m_1^i \\
&\vdots \\
-\left(\frac{k_1^i}{(p_1^i)^q} + \frac{k_2^i}{(p_2^i)^q} + \ldots + \frac{k_q^i}{(p_q^i)^q}\right) &= m_{q-1}^i
\end{aligned}
\tag{15}
$$

This may be rewritten in matrix form [20] as

$$\vec{k}^i = -\mathcal{V}^{-1} \vec{m}_L^i \tag{16}$$

where $\vec{m}_L^i$ is a vector of the *low-order* coefficients, $(m_0^i, m_1^i, \ldots, m_{q-1}^i)^T$, and $\mathcal{V}$ is the matrix

$$
\begin{bmatrix}
(p_1^i)^{-1} & (p_2^i)^{-1} & \cdots & (p_q^i)^{-1} \\
(p_1^i)^{-2} & (p_2^i)^{-2} & \cdots & (p_q^i)^{-2} \\
\vdots & \vdots & \ddots & \vdots \\
(p_1^i)^{-q} & (p_2^i)^{-q} & \cdots & (p_q^i)^{-q}
\end{bmatrix}
$$

4. Order of approximation. The automatic selection of an appropriate order of approximation q is a very hard problem, and may be influenced by several factors: the required approximation accuracy, circuit behavior, numerical precision, and signal bandwidth. Previous work suggested a stopping criterion based on the results of successive orders of approximation [19]. A normalized root-mean-square error was measured between the qth and $(q + 1)$th orders of approximation; when this error decreased below a particular value, the approximation process was stopped.

However, it would be more prudent to use a stopping criterion based on a combination of several of the factors mentioned above. In [13], Huang showed that as the order of approximation was increased, the model poles converged to actual system poles. For a particular signal bandwidth $fmax$, it would therefore seem necessary to increase the order of approximation until those poles corresponding to all frequencies below and near $fmax$ have converged.

For signals with very small rise-times, this approach might lead to excessively high orders of approximation that might not be realizable by a stable model, given the finite numerical precision used. This effect of numerical precision refers to the increasing sensitivity of the model simplification scheme to errors in the series coefficient values as the order of approximation is increased. By examining the coefficient values, a practical limit on the highest order possible with a given set of coefficients, can be derived as shown below. The power series representation of $[\vec{X}(s)]_i$, from Eq.(8)-(10), is

$$[\vec{X}(s)]_i = m_0^i + m_1^i s + m_2^i s^2 + \cdots \tag{17}$$

The Hankel matrix representing this power series for the first $(2q - 1)$ coefficients is

$$H_q^{(0)} = \begin{bmatrix} m_0^i & m_1^i & \cdots & m_{q-1}^i \\ m_1^i & m_2^i & \cdots & m_q^i \\ \vdots & \vdots & \ddots & \vdots \\ m_{q-1}^i & m_q^i & \cdots & m_{2q-2}^i \end{bmatrix} \tag{18}$$

This is also recognized to be the matrix in Eq.(13) for the roots of the characteristic polynomial.

From [12, 5], it is seen that the degree of a proper rational function is equal to the rank of the Hankel matrix representing its power series expansion. In the development of $[\vec{X}(s)]_i$, which is indeed a proper rational-function approximation, the degree would represent the order of approximation that is sought, i.e., the degree of $[\vec{X}(s)]_i$. Further, since, by assumption, the series expansions of the actual response and the reduced-order model are to agree at least as far as the first $2q$ coefficients, Equation (18) would represent the series expansion of $[\hat{X}(s)]_i$ as well. Hence, an upper limit on the order of approximation would be $q \leq \rho H_q^{(0)}$ where ρ denotes the rank. Attempting to obtain an order of approximation higher than this limit would result in the truncation noise being personified by unstable poles or poles with relatively insignificant residues. (A similar approach was also suggested to us by Pak Chan [4]).

Thus, it might not be possible to fit the signal bandwidth with a set of poles spanning that bandwidth. Rather, the order of approximation is increased until either the bandwidth requirement is satisfied, or there is amplification of numerical noise in the form of poles with vanishing residues.

As an example, consider the 4000-node RC-tree in Fig.2. The approximate domi-

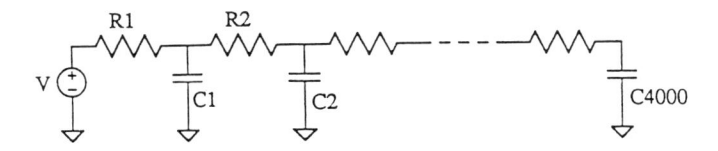

FIG. 2. *4000-node RC-tree circuit.*

nant poles and residues of the response at the output node are shown in Table 1 for increasing orders of approximation. To obtain an approximation for a signal band-

TABLE 1
Poles and residues at output of 4000-node RC tree for increasing orders of approximation.

Order	Poles	Residues
1	-5.00467e+07	-1.00000e+00
2	-6.17908e+07	-1.27681e+00
	-4.05788e+08	2.76814e-01
3	-6.17363e+07	-1.27303e+00
	-5.75856e+08	5.03070e-01
	-9.80745e+08	-2.30044e-01
4	-6.17363e+07	-1.27303e+00
	-5.57263e+08	4.25811e-01
	-1.41178e+09	-1.90165e-01
	-3.86961e+09	3.73820e-02
5	-6.17363e+07	-1.27303e+00
	-5.57018e+08	4.24933e-01
	-1.54068e+09	-2.52229e-01
	-2.66287e+09	1.06412e-01
	-1.41686e+11	-6.08759e-03
6	-6.17363e+07	-1.27303e+00
	-5.57017e+08	4.24892e-01
	-1.54286e+09	-2.51642e-01
	-2.70264e+09	1.06551e-01
	-1.45949e+11	-6.77357e-03
	-5.73723e+07	-9.54016e-09
7	-6.17363e+07	-1.27303e+00
	-5.57110e+08	4.25406e-01
	-1.50857e+09	-2.38670e-01
	-2.68746e+09	9.06235e-02
	-1.45735e+11	-4.33107e-03
	-5.37956e+07	3.19245e-10
	-1.62724e+07	-5.20880e-17

width of, say, 5e+8 radians, a 3rd-order approximation would be sufficient, since all the poles below that frequency have converged and do not shift appreciably at higher orders. However, for signal frequencies much greater than 1e+11, attempting to obtain models of order greater then 5 would yield poles with relatively insignificant residues that do not influence the response. These poles would represent the magnification of numerical noise and occur at random locations, as illustrated by the "noise" poles in the 6th- and 7th-order approximations in Table 1.

The next section introduces concerns of instability that are inherent to moment-matching and Padé approximation techniques, in particular. These concerns play an important part in the actual approximation process.

5. Moment-matching instability. Theorem 2.1 requires the asymptotic stability of the Padé approximant $\mathcal{L}\{\hat{g}\}$. However, obtaining the model poles from Eq.(12)-(14) may yield unstable models of systems that are asymptotically stable. In the case of stable, linear(ized) RLC circuits, this instability manifests itself in the form of unstable model poles, or poles in the right-half of the s-plane. This is due to an inherent instability problem associated with the Padé approximation [3] and moment-matching methods in general.

As explained by Huang [13], the Zinn-Justin theorem [27] on the convergence of the Padé approximants demonstrates the uniform convergence of the diagonal Padé sequence to the actual system function, except in *exceptional* areas of the complex plane. These exceptional areas can be made arbitrarily small by increasing the order of approximation. This leads to the implication that those model poles that correspond to actual system poles appear repeatedly in increasing orders of approximation [13]. This phenomenon is seen in the results in Table 1 on page 7, where the convergence of the minimum poles is clearly observed.

The model poles that correspond to actual system poles are easily distinguished by their repeated occurrence and significant residues. However, those model poles that do not correspond to any system pole, termed *defective* poles [13], are distinguished by their insignificant residues and random occurrence. These defective poles may occur at any order of approximation due to their random nature, and may even cause the model to become unstable.

One of the reasons for the occurrence of these defective poles is the extreme sensitivity of the Padé approximation technique to errors in the coefficient values. Errors are due to truncation/rounding when working with finite-precision machines. Table 2 demonstrates this extreme sensitivity for the response at node 2000 of a 4000-node RC-tree. A change of 1e-57 in the value of the 8th moment causes a stable fourth-order approximation to become unstable.

In addition to numerical noise, certain system-pole patterns resist approximation at certain orders. Using another example from [13], the pole-pattern in Fig.3 depicts a system function with several complex pole-pairs and numerous other insignificant poles. Attempting an approximation with an odd number of model poles will yield a model pole that does not correspond to any system pole, and may result in instability.

Similarly, the locations of the system zeros play a significant part in the stability of the reduced-order model. This can be illustrated using the contrived 4-pole system in Table 3. For this example, the system poles were fixed while the locations of the zeros were varied, as in [13], and the stability of the resultant models at different orders

TABLE 2
Illustration of sensitivity of moment-matching scheme to noise in the coefficient values of response at node 2000 of 4000-node RC tree (perturbed coefficient in box).

Actual coefficients	poles	residues
1.000000000000000e+00	-6.17363e+07	-8.99251e-01
-1.497267448118228e-08	-5.56846e+08	-3.01767e-01
2.368287909248411e-16	-1.66283e+09	2.32082e-01
-3.823417609189066e-24	-7.18473e+09	-3.10641e-02
6.190703522181460e-32		
-1.002720592080674e-39		
1.624190893290490e-47		
$\boxed{-2.630850265524870\text{e-}55}$		
Perturbed coefficients	poles	residues
1.000000000000000e+00	-6.17363e+07	-8.99254e-01
-1.497267448118228e-08	-5.72767e+08	-3.39101e-01
2.368287909248411e-16	-1.28563e+09	2.38355e-01
-3.823417609189066e-24	3.73120e+04	-7.39886e-21
6.190703522181460e-32		
-1.002720592080674e-39		
1.624190893290490e-47		
$\boxed{-2.629163340894621\text{e-}55}$		

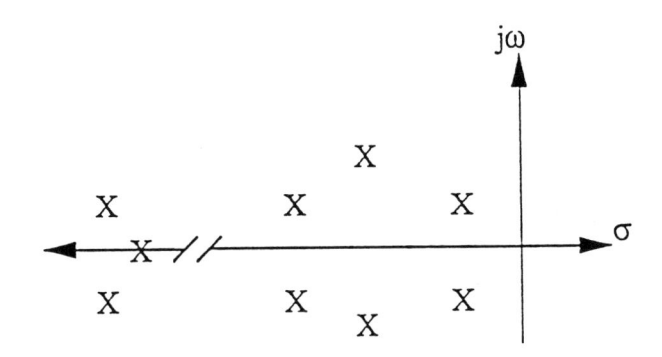

FIG. 3. *Pole-pattern of an artificial system function.*

TABLE 3
Effect of the location of system zeros on the stability of the reduced-order models of a 4-pole, 3-zero function.

	System poles	-1	-10	-15	-100
Case 1	System zeros	-2	-20	-40	
	O(1)-model poles	-1.66			
	O(2)-model poles	-1.04	-22.31		
	O(3)-model poles	-1.00	-9.23	-158.87	
Case 2	System zeros	-2	-13	+6	
	O(1)-model poles	-1.30			
	O(2)-model poles	-1.16	*+895.36*		
	O(3)-model poles	-1.00	-11.08	-104.44	
Case 3	System zeros	+19	+8	-6	
	O(1)-model poles	-0.84			
	O(2)-model poles	-1.00	*+84.12*		
	O(3)-model poles	-1.00	*+3.61*	*+74.63*	

of approximation was observed. A better understanding of all the aforementioned effects would greatly aid in the development of techniques to improve the reliability of moment-matching methods.

The problem of instability associated with moment-matching methods has been the focus of numerous papers in several branches of engineering. An approach suggested by Brown [3] to remedy the instability of a reduced-order model of a stable system was to increase the order of approximation, i.e., increase the number of dominant poles in the model. An auxiliary performance criterion was also proposed for computing the additional parameters of the model in the event of the non-availability of the extra system-moments required. However, caution must be exercised not to exceed the order of the actual system itself.

In contrast, Zakian [26] proposes successively decreasing the order of approximation until a stable model is obtained. While a stable model may be eventually achieved, the order at which stability is achieved may be too low for the application, resulting in loss of accuracy. However, this method is also not fool-proof and it is possible to obtain transfer functions with unstable *1st-order* models [13].

Besides the above two, numerous other techniques have been proposed for ensuring model-stability in moment-matching approximations [8, 7, 9, 15, 18, 24, 25]. However, most of these techniques are unsuitable for circuit analysis. In the following section, some recent approaches to overcoming the instability problem as applied to electrical networks is described.

6. Minimizing instability. While the inherent instability of the Padé approximation is difficult to detect or remedy without *a priori* knowledge of the actual transfer function, numerical instability can be minimized. One of the most apparent problems stems from the rapid divergence of the coefficients of the series expansion of the actual response, as seen in Table 2. Such widely varying magnitudes of the coefficients may cause the matrix in Eq.(13) to tend to singularity at very low orders of approximation.

This ill-conditioning of the coefficient matrix can be minimized by employing frequency scaling of the coefficient values. The normalized coefficients are used to find a normalized solution which can be scaled back to obtain the desired values. In AWE, the scale factor selected, with respect to the impulse-response coefficients $\vec{m}$, is:

$$\gamma = \frac{m_0}{m_1}. \tag{19}$$

The normalization is achieved by scaling the jth impulse-response coefficient by γ^j. The scaled coefficients are used to find a set of normalized poles and residues. The desired values are then recovered by scaling back the poles by γ. An example of this scaling technique is shown in Table 4.

TABLE 4
Frequency scaling of coefficient values of response at output of 4000-node RC-tree driven by a 5V step excitation.

coefficient	unscaled	scaled
m_0	5.000000000000000e+00	5.000000000000000e+00
m_1	-9.990664920421377e-08	-5.000000000000000e+00
m_2	1.663647067197849e-15	4.166893693654019e+00
m_3	-2.703917792898261e-23	-3.389380423777905e+00
m_4	4.381506365320577e-31	2.748690814396173e+00
⋮	⋮	⋮
Scale factor γ: 5.004671901046116e+07		

However, in some cases, using the scaled coefficient values may still result in the near-singularity of the Hankel matrix in Eq.(18). This was demonstrated by Huang [13] who showed that the higher-order coefficients are increasingly influenced by the minimal poles. Further, the contribution of the high-frequency poles decreases with the ratio of the magnitudes of the low- and high-frequency poles. The effect of scaling is not evinced here, since the *ratio* of the magnitudes of the poles is unaffected.

To overcome this, a method of frequency shifting, termed *uniform predistortion*, was suggested in [13]. As compared to frequency scaling, where the energy-storage elements are scaled, frequency shifting involves adding proportional resistors in parallel or series to the energy storage elements. This has the effect of moving the $j\omega$-axis to the right, as shown in Fig.4, and thus increasing the ratio of low- to high-frequency poles. With respect to Fig.4,

$$\frac{p_1 + \lambda}{p_2 + \lambda} > \frac{p_1}{p_2}.$$

The degree of shift, λ, of the $j\omega$ axis determines the change in the ratio of pole magnitudes.

Another technique of overcoming numerical instability, for the special case of passive, linear RC-interconnect circuits, is described in [11, 10]. This approach attempts to map a set of series coefficients of the homogeneous step response, to a stable dominant pole-residue representation using constrained optimization. This is facilitated by

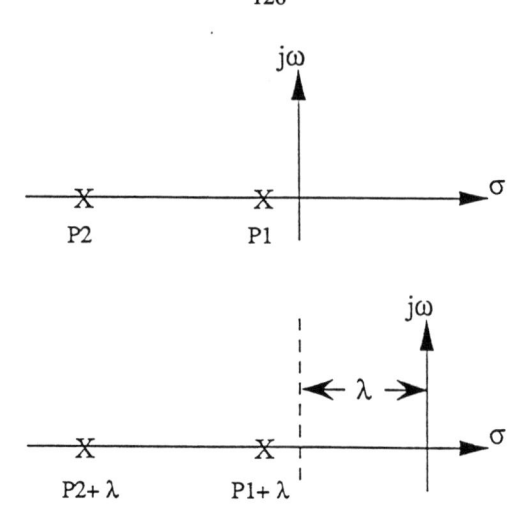

FIG. 4. *Frequency shifting to improve the ratio of pole magnitudes.*

the *a priori* knowledge that the poles of passive, linear RC-circuits are real and negative. Hence, for a stable approximation, the model-poles should be real and negative as well. This forms a nonlinear inequality constraint that can be easily incorporated in the form of a variable transformation, $[p_j = -exp(x_j)]$ on the system in Eq.(15). The resultant, transformed system is optimized in x-space using unconstrained techniques.

This constrained optimization technique is employed in RICE (Rapid Interconnect Circuit Evaluator) [23], an implementation of AWE for the analysis of interconnect circuits. RICE uses an efficient path-tracing scheme [22], that minimizes introduction of numerical errors, for the computation of the moments of the circuit responses. In addition, problem conditioning is maintained throughout through the use of frequency scaling and the use of numerical techniques such as the singular value decomposition [11, 10].

Table 5 shows the results of using RICE to model the step response at node 1000 of the RC-tree in Fig.2. A 3rd-order unconstrained approximation yields an unstable

TABLE 5
Unconstrained and constrained models of the step response at node 1000 of 4000-node RC-tree.

poles (p_l)		residues (k_l)	
Unoptimized	Optimized	Unoptimized	Optimized
-6.17370e+07	-6.17274e+07	-4.90604e-01	-4.90336e-01
-6.08460e+08	-5.22779e+08	-5.13784e-01	-5.51879e-01
5.49704e+08	-2.37423e+08	4.38880e-03	4.26101e-02

model, while using the constrained optimization scheme yielded a stable model which compares very favorably with the output of a circuit simulator [21], as shown in Fig.5.

The RICE software represents an application-specific implementation of AWE that exploits the tree-like topology of interconnect circuits. However, even a generalized version of AWE [14], when compared with a circuit simulator [21], displays the

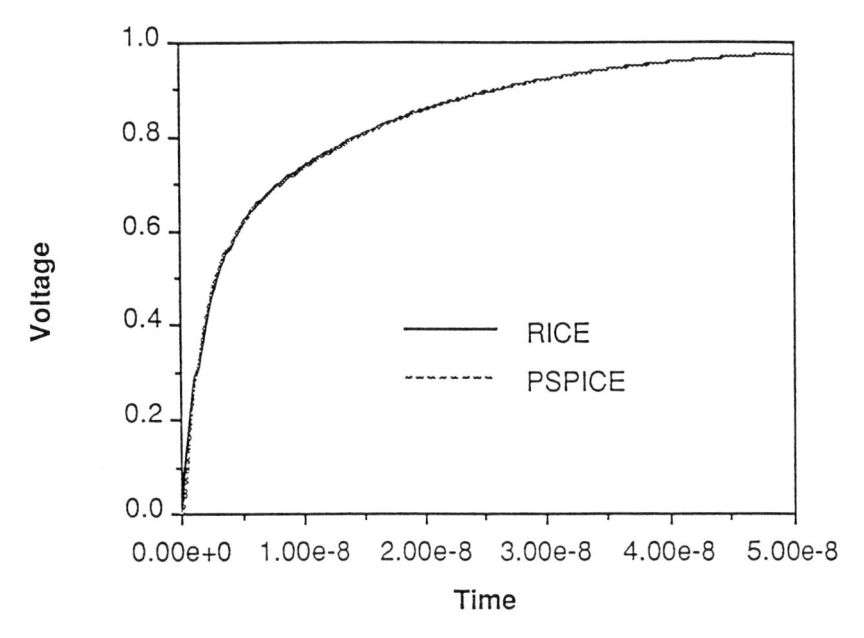

FIG. 5. *Comparison of the constrained 3rd-order AWE model of the response of a 4000-node RC-tree versus the output of a circuit simulator.*

TABLE 6

*Execution time (in seconds) for various sizes of RC-interconnect circuit models. (An * indicates the circuit was too large in terms of memory requirement.)*

RC-Interconnect circuit size				
Branches	Nodes	RICE	AWE(LU)	PSPICE
4,001	1,601	0.07	5.79	97.6
16,001	6,401	0.28	57.43	908.7
64,001	25,601	1.17	921.7	*

tremendous improvements in performance that motivate the use of moment-matching methods for certain applications (Refer Table 6).

Although the results reported in Table 6 are for passive RC-interconnect circuits, the implementation in RICE works equally efficiently for passive RLC-interconnect models, as shown in Fig.6. The poles and residues of passive RLC circuits may be complex, although they are still located in the left-half of the complex plane. Since these circuits have a much higher response bandwidth than the RC circuits discussed earlier, a much higher order of approximation is required to obtain a response model that matches the output of a circuit simulator. Consequently, the instability problem is more pronounced and occurs more frequently in RLC-circuit response models than in RC-circuit response models. However, the gains in computation time to be achieved by using AWE rather than a circuit simulator, are also multiplied by several orders of magnitude. Fig.7 shows the comparison of the step response of the circuit in Fig.6

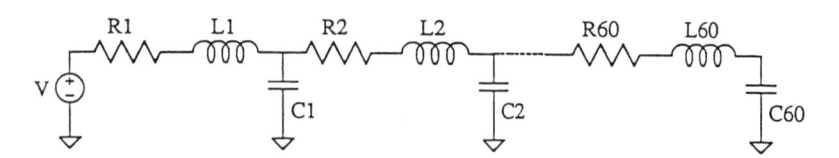

FIG. 6. *Typical RLC-interconnect circuit model.*

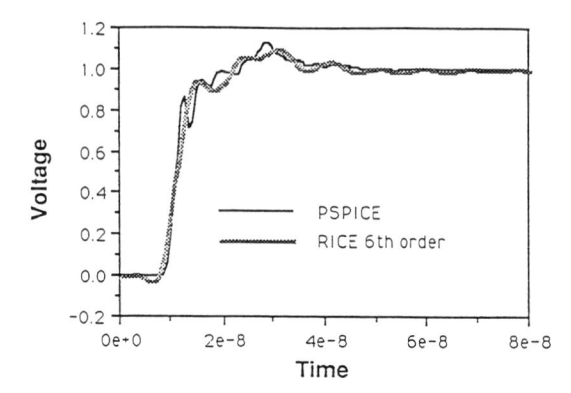

FIG. 7. *Comparison of the 6th-order AWE model of the step response of an RLC-interconnect circuit versus the output of a circuit simulator.*

obtained from PSPICE and a 6th-order AWE model. Note that even a 6th order model does not capture all of the high frequency effects due to an ideal step input. As the rise time of the input increases, the AWE model approximates the actual waveform more closely. Fig.8 shows this observation. The greater the rise time of the input signal, the lesser is it's high frequency content, hence, the lower the required order of approximation.

Circuits with controlled-sources and active devices may be inherently asymptotically unstable and may possess transfer-function poles in the right half of the complex plane. In such cases, AWE reduces to a "pure" Padé approximation, since the moments of the response of a function with a positive pole cannot be obtained due to the divergent nature of the response [16]. However, the problem posed here is the determination of whether a positive model-pole reflects the instability problem associated with the Padé approximation or is an approximation to the actual positive system-pole.

7. Conclusion. Asymptotic Waveform Evaluation has been demonstrated as an efficient approach to waveform estimation for linear RLC circuits, interconnects in particular. Indeed, RICE, which is an application-specific implementation of AWE, has proven to be capable of analyzing stiff interconnect-circuit models several orders of magnitude faster than a circuit simulation, with no loss in accuracy.

However, there remain many unanswered questions regarding moment-matching and Padé approximation, some of which have eluded eminent researchers since a long

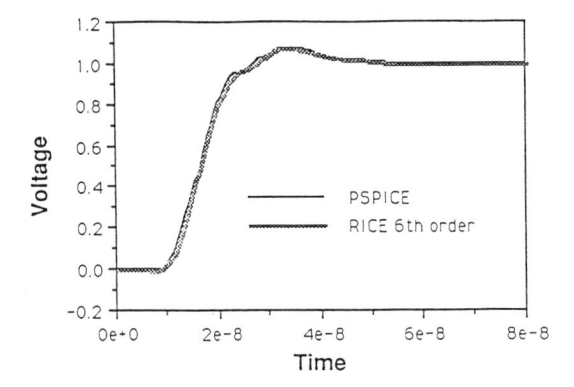

FIG. 8. *Comparison of the 6th-order AWE model and the output of a circuit simulator for a 10ns input-signal rise time.*

time. Alleviating these problems of instability and order-estimation will enable the extension of AWE to even more complex and challenging tasks.

Acknowledgements. The authors would like to thank Demosthenes F. Anastasakis for his helpful discussions and assistance in the preparation of this document.

REFERENCES

[1] G. A. Baker, Jr. *Essentials of Pade Approximants.* Academic Press, 1975.
[2] G. A. Baker, Jr. and P. Graves-Morris. *Encyclopedia of Mathematics and its Applications,* volume 13. Addison-Wesley Publishing Co., 1981.
[3] R. F. Brown. Model stability in use of moments to estimate pulse transfer functions. *Electron. Lett.,* 7, 1971.
[4] P. K. Chan. Comments on asymptotic waveform evaluation for timing analysis. Private correspondence.
[5] C. Chen. *Linear System Theory and Design.* CBS Collge Publishing, 1984.
[6] L. O. Chua and P. Lin. *Computer-Aided Analysis of Electronic Circuits: Algorithms and Computational Techniques.* Prentice-Hall, Inc., 1975.
[7] E. J. Davison. A method for simplifying linear dynamic systems. *IEEE Trans. Auto. Control,* 11, Jan 1966.
[8] J. F. J. Alexandro. Stable partial Pade' approximations for reduced-order transfer functions. *IEEE Trans. Auto. Control,* 29, 1984.
[9] L. G. Gibilaro and F. P. Lees. The reduction of complex transfer function models to simple models using the method of moments. *Chem. Eng. Sc.,* 24, 1969.
[10] N. Gopal and L. T. Pillage. Constrained approximation of dominant time constant(s) in RC circuit delay models. Technical Report TR-CERC-TR-LTP-91-01, Comp. Eng. Res. Ctr., U. Texas (Austin), Jan 1991.
[11] N. Gopal, C. Ratzlaff, and L. T. Pillage. Constrained approximation of dominant time constant(s) in RC circuit delay models. In *Proc. 13th IMACS World Congress Comp. App. Math.,* Jul 1991.
[12] P. Henrici. *Applied and Computational Complex Analysis.* John Wiley & Sons, 1974.
[13] X. Huang. *Pade' approximation of linear(ized) circuit responses.* PhD thesis, Carnegie Mellon Univ., Nov 1990.
[14] X. Huang, V. Raghavan, and R. A. Rohrer. AWEsim: A program for the efficient analysis of linear(ized) circuits. In *Proc. IEEE Int'l. Conf. Computer-Aided Des.,* Nov 1990.
[15] M. F. Hutton and B. Friedland. Routh approximations for reducing order of linear time-invariant systems. *IEEE Trans. Auto. Control,* 20, 1975.
[16] S. M. Kendall and A. Stuart. *The Advanced Theory of Statistics.* MacMillan Pub. Co., Inc., 1977.

[17] S. P. McCormick. *Modeling and Simulation of VLSI Interconnections with Moments.* PhD thesis, Mass. Inst. Tech., June 1989.

[18] J. Pal. Stable reduced-order Pade' approximants using the Routh-Hurwitz array. *Electron. Lett.*, 15, 1979.

[19] L. T. Pillage. *Asymptotic Waveform Evaluation for Timing Analysis.* PhD thesis, Carnegie Mellon Univ., Apr 1989.

[20] L. T. Pillage and R. A. Rohrer. Asymptotic waveform evaluation for timing analysis. *IEEE Trans. Comp. Aided Design*, 9, 1990.

[21] PSPICE USER'S MANUAL. *Version 4.03.* Microsim Corp., Jan 1990.

[22] C. L. Ratzlaff. A fast algorithm for computing the time moments of RLC circuits. Master's thesis, The Univ. of Texas at Austin, May 1991.

[23] C. L. Ratzlaff, N. Gopal, and L. T. Pillage. RICE: Rapid Interconnect Circuit Evaluator. In *Proc. 28th ACM/IEEE Design Auto. Conf.*, Jun 1991.

[24] R. H. Rosen and L. Lapidus. Minimum realization and systems modeling: Part I - Fundamental theory and algorithms. *A. I. Ch. E. J .*, 18, Jul 1972.

[25] Y. Shamash. Linear system reduction using Pade approximation to allow retention of dominant modes. *Int'l. J. Control*, 21(2), 1975.

[26] V. Zakian. Simplification of linear time-invariant systems by moment approximants. *Int'l. J. Control*, 18, 1973.

[27] J. Zinn-Justin. Strong interaction dynamics with Pade' approximants. *Phy. Rep.*, 1970.

SPECTRAL ALGORITHM FOR SIMULATION OF INTEGRATED CIRCUITS

O.A. PALUSINSKI, F. SZIDAROVSZKY, C. MARCJAN, AND M. ABDENNADHER[*],

Abstract. Waveform relaxation improves the efficiency of integrated circuits transient simulation at the expense of large memory needed for storage of coupling variables and complicated intersubcircuit communication requiring interpolation. A new integration method based on the expansion of unknown variables in Chebyshev series is developed. Such a method assures very compact representation of waveforms, minimizing storage requirements. Solutions are provided in continuous form, therefore no extra interpolation is needed in the iterations. The resulting algorithm was implemented and proved to be very efficient. A short description of spectral technique is presented and an application of spectral analysis in computing the transient behavior of an MOS circuit is discussed. The computing proved to be much more efficient in comparison with other methods.

1. Introduction. Electronic circuits are mathematically represented by a set of nonlinear differential equations formulated using modified nodal analysis (MNA). A method based on Newton–Kantorovich's approach is used for linearization of the circuit. The resulting linearized system is in the form of a set of first order differential equations which are solved with application of spectral analysis. Spectral technique framework based on Chebyshev polynomials with their properties is applied in the prototype software (SPEC) yielding accurate (globally controllable accuracy) and fast [1] simulation processes.

2. Representation of functions using Chebyshev series. The expansion of a function c(t) defined in the interval [-1, 1] is written as an infinite series as follows

$$(2.1) \qquad c(t) = \sum_{i=0}^{\infty}{}' c_i T_i(t),$$

where $T_i(t)$ denotes a first kind Chebyshev polynomial of ith degree and c_i are the constant coefficients [2]. The prime at the summation symbol denotes that the first term in the summation is halved. If a function $\tilde{c}(\tilde{t})$ is defined in a general interval $[\tilde{t}_1, \tilde{t}_2]$ then it has to be scaled to the interval [-1, 1] to obtain the scaled function $c(t)$ used in (2.1). The scaling is performed using the following operation

$$(2.2) \qquad t = \frac{2\tilde{t} - \tilde{t}_1 - \tilde{t}_2}{\tilde{t}_2 - \tilde{t}_1}.$$

To simplify notation the equation (2.1) can be rewritten in the following vector form

$$(2.3) \qquad \mathcal{C}\{c(t)\} = \mathbf{c},$$

---

[*] Department of Electrical and Computer Engineering, University of Arizona, Tucson, AZ 85721

where $\mathcal{C}$ denotes the Chebyshev transformation and the entries of the vector **c** are composed of expansion coefficients of the function $c(t)$ as

$$(2.4) \qquad \mathbf{c} = \begin{pmatrix} c_0 \\ c_1 \\ c_2 \\ \vdots \end{pmatrix}.$$

Using properties of Chebyshev polynomials, transformation of the product of two functions $c(t)$ and $g(t)$ can be written as

$$(2.5) \qquad \mathcal{C}\{c(t)g(t)\} = \mathbf{Cg} = \mathbf{Gc},$$

where

$$(2.6a) \qquad \mathbf{C} = \begin{pmatrix} \frac{1}{2}c_0 & c_1 & \cdots & & c_j & \cdots \\ \frac{1}{2}c_1 & \ddots & & & & \\ \vdots & & & & & \\ \frac{1}{2}c_i & & & \frac{1}{2}(c_{i+j} + c_{|i-j|}) & & \\ \vdots & & & & & \end{pmatrix}$$

and

$$(2.6b) \qquad \mathbf{G} = \begin{pmatrix} \frac{1}{2}g_0 & g_1 & \cdots & & g_j & \cdots \\ \frac{1}{2}g_1 & \ddots & & & & \\ \vdots & & & & & \\ \frac{1}{2}g_i & & & \frac{1}{2}(g_{i+j} + g_{|i-j|}) & & \\ \vdots & & & & & \end{pmatrix}.$$

Chebyshev series offer a simple relation between the coefficients of expansion of a function $y(t)$ and the coefficients of expansion of its derivative $y'(t)$ given by

$$(2.7) \qquad \mathbf{y} = \mathbf{By}^* + 2\mathbf{e}y(-1)$$

where

$$(2.8) \qquad \mathbf{y}^* = \mathcal{C}\{y'(t)\},$$

(2.9) $$\mathbf{e} = \begin{pmatrix} 1 \\ 0 \\ 0 \\ \vdots \end{pmatrix},$$

and the matrix $\mathbf{B}$ is invariant and its first row is defined as:

(2.10) $$B_{0i} = (\tilde{t}_2 - \tilde{t}_1) \cdot \begin{cases} \frac{1}{2} & ; \quad i = 0 \\ -\frac{1}{4} & ; \quad i = 1 \\ \frac{(-1)^{i+1}}{(i-1)(i+1)} & ; \quad i = 2, 3, \ldots \end{cases}$$

and the remaining entries are

(2.11) $$B_{ij} = (\tilde{t}_2 - \tilde{t}_1) \cdot \begin{cases} \frac{1}{4i} & ; \quad j = i - 1 \\ -\frac{1}{4i} & ; \quad j = i + 1 \\ 0 & ; \quad \text{elsewhere.} \end{cases}$$

3. Solving differential equations using Chebyshev series. Linear differential equations are easily solved using spectral techniques with the use of Chebyshev series and their properties. In this section a numerical algorithm is presented. Consider the following scalar differential equation

(3.1) $$\tilde{c}(\tilde{t})\frac{d\tilde{y}}{d\tilde{t}} = \tilde{g}(\tilde{t})\tilde{y} + \tilde{h}(\tilde{t})$$

(3.1a) $$\tilde{y}(\tilde{t}_1) = y_0,$$

where functions $\tilde{c}(\tilde{t})$, $\tilde{g}(\tilde{t})$ and $\tilde{h}(\tilde{t})$ are defined in the time interval

(3.2) $$\tilde{t}_1 \le \tilde{t} \le \tilde{t}_2,$$

and $\tilde{y} = \tilde{y}(t)$ is the unknown function on the same interval. To simplify mathematical operations equation (3.1) is scaled to the interval $[-1, 1]$ as described in the previous section. As a result, equation (3.1) is rewritten as

(3.3) $$c(t)\frac{dy}{dt} = g(t)y + h(t)$$

(3.3a) $$y(-1) = y_0,$$

where $y = y(t)$ is the unknown function, and $c(t)$, $g(t)$ and $h(t)$ are known functions of t on the interval $[-1, 1]$. Using the notation introduced in the previous section the above equation can be rewritten in the following transformed form as a relation between the Chebyshev expansion coefficients

$$\mathbf{Cy^* = Gy + h}. \tag{3.4}$$

Using relation (2.7) the above equation is rearranged in the following form

$$(\mathbf{C - GB})\mathbf{y^*} = \tilde{y}(\tilde{t}_1)\mathbf{g + h}, \tag{3.5}$$

which can be easily solved since it is a set of linear algebraic equations, where matrices **C, G, B** and vectors **g** and **h** are known, defined in section 2.

In practice, Chebyshev expansions of the functions are performed to some selected degree which depends on the required accuracy of the solution [3]. When the solution is of order N, then analyzing equation (3.4) and relations (2.6ab), the forcing function $h(t)$ is expanded into Chebyshev series of degree N, however functions $c(t)$ and $g(t)$ are of degree $2N$.

4. Simulation of MOS circuits. Analyzed MOS circuits are described as a set of M nonlinear differential equations with M unknown functions formulated by using MNA [4]. Those equations are scaled and rewritten in the following form:

$$\mathbf{C}(\mathbf{v}(t), t)\frac{d\mathbf{v}(t)}{dt} = \mathbf{q}(\mathbf{v}(t), t) \qquad \mathbf{v}(-1) = \mathbf{v_0}, \tag{4.1}$$

where $\mathbf{v}(t)$ is an M-dimensional vector of circuit variables, $\mathbf{v_0}$ is a vector of initial conditions, **C** is an $M \times M$ square matrix with variable entries and **q** is a M-dimensional vector of nonlinear functions.

Special Case

Consider the scalar case ($M = 1$), linearize equation (4.1) around a given waveform $y_p(t)$ to yield

$$c^p(t)\frac{dv}{dt} = g^p(t)v(t) + h^p(t) \qquad v(-1) = y_0 = v_0, \tag{4.2}$$

where

$$\begin{aligned} c^p(t) &= c(y_p(t), t), \\ g^p(t) &= \left.\frac{\partial q}{\partial v}\right|_{v=y_p(t)} - \left.\frac{\partial c}{\partial v}\right|_{v=y_p(t)} \frac{dy_p}{dt}, \\ h^p(t) &= q[y_p(t), t] - g^p(t)y_p(t). \end{aligned} \tag{4.3}$$

Equation (4.2) is solved by using the same procedure as in the case of equation (3.3). Based on this linearization a Newton-Kantorovich-type iteration procedure is used to recover the solution. The process continues until the iterates satisfy a convergence condition

(4.4) $$\| v^{k+1}(t) - v^k(t) \|_\infty \leq \epsilon,$$

where ϵ is the convergence tolerance, k denotes the iteration count, and $\| \cdot \|_\infty$ denotes the maximum-norm.

General Case

Equation (4.1) can be rewritten as

(4.5) $$\begin{pmatrix} c_{11} & c_{12} & \cdots & c_{1M} \\ c_{21} & c_{22} & \cdots & c_{2M} \\ \vdots & \vdots & & \vdots \\ c_{M1} & c_{M2} & \cdots & c_{MM} \end{pmatrix} \frac{d}{dt} \begin{pmatrix} v_1 \\ v_2 \\ \vdots \\ v_M \end{pmatrix} = \begin{pmatrix} q_1 \\ q_2 \\ \vdots \\ q_M \end{pmatrix},$$

where

(4.6a) $$c_{ij} = c_{ij}(\mathbf{v}, t)$$

(4.6b) $$q_i = q_i(\mathbf{v}, t).$$

Each subcircuit is linearized around a vector of waveform

(4.7) $$\mathbf{y_p} = \begin{pmatrix} v_{p0}(t) \\ v_{p1}(t) \\ \vdots \\ v_{pM}(t) \end{pmatrix}$$

yielding an equation of ith subcircuit:

(4.8) $$\sum_{j=1}^{M} c_{ij}(\mathbf{y_p}, t) \frac{dv_j}{dt} = \sum_{j=1}^{M} g_{ij}(t) v_j + q_i(\mathbf{y_p}, t) - \sum_{j=1}^{M} g_{ij}(t) v_{pj},$$

where

(4.9) $$g_{ij}(t) = \frac{\partial q_i}{\partial v_j}\bigg|_{v=v_p} - \sum_{k=1}^{M} \frac{\partial c_{ij}}{\partial v_k}\bigg|_{v=v_p} \frac{dv_{pj}}{dt}.$$

After linearizing all subcircuits using (4.8) and (4.9), equation (4.7) is rewritten in the form:

(4.10) $$\mathbf{C}^p(t)\frac{d\mathbf{v}}{dt} = \mathbf{G}^p(t)\mathbf{v} + \mathbf{h}^p(t),$$

where $\mathbf{C}^p(t)$, $\mathbf{G}^p(t)$ are $M \times M$ matrices and $\frac{d\mathbf{v}}{dt}$, $\mathbf{v}$ and $\mathbf{h}^p(t)$ are M dimensional vectors. Each element of matrices $\mathbf{C}^p(t)$, $\mathbf{G}^p(t)$ is expanded into Chebyshev series of degree $2N$ and formes an $(N+1) \times (N+1)$ submatrix defined by (2.6ab). The elements of the vector $\mathbf{h}^p(t)$ are expanded into Chebyshev series of degree N and form subvectors of order $(N+1)$ of coefficients of expansion. Using property (2.7) the constructed equation for expansion coefficients can be assembled and represented in a matrix form as:

(4.11) $$(\mathcal{A}^p - \mathcal{G}^p \mathcal{B})\vartheta^* = \mathcal{G}^p \mu + \chi,$$

where $\mathcal{A}^p$, $\mathcal{G}^p$, $\mathcal{B}$ are $M(N+1) \times M(N+1)$ square matrices and μ, χ are $M(N+1)$ dimensional vectors. The vector ϑ^* is composed of $(N+1)$ dimensional subvectors $\mathbf{v}_j^*$. A subvector $\mathbf{v}_j^*$ contains the coefficients of the expansion of the jth component of $\frac{d\mathbf{v}}{dt}$. The blocks of matrices $\mathcal{A}^p$, $\mathcal{G}^p$ are determined by the submatrices created from expansion of $\mathbf{C}^p(t)$ and $\mathbf{G}^p(t)$. Matrix $\mathcal{B}$ is a block diagonal with blocks composed of M identical $(N+1) \times (N+1)$ square matrices defined by relations (2.10) and (2.11). Vector χ is composed of $(N+1)$ dimensional subvectors each represents expansion coefficients of a respective element of vector $\mathbf{h}^p(t)$. Vector μ is composed of the $(N+1)$ dimensional subvectors

(4.12) $$\mu_k = 2v_{0k}\mathbf{e}$$

$$k = 1, 2, ..., M,$$

which depends on the initial conditions v_{0k} for the kth components of vector $\mathbf{v}$. The $M(N+1)$ dimensional vector ϑ is calculated by using the relation

(4.13) $$\vartheta = \mathcal{B}\vartheta^* + \mu.$$

The $M(N+1)$ vector ϑ is composed of M subvectors $\mathbf{v_k}$ each containing $(N+1)$ expansion coefficients.

5. Example.

5.1. Circuit model. Simulation of MOS circuits is based on the solution of ordinary differential equations obtained using MNA. The circuit equations are written in the general form (4.1) In order to provide better description of the model and solution algorithm details a specific example is given below. Model of CMOS NAND Gate The schematic of NAND gate built in CMOS technology is shown in Fig. 1. The equivalent circuit is obtained by replacing the MOS transistors by appropriate model [5] and the equations for the unknown nodal voltages (V_3, V_4) are written in the matrix form

$$(5.1) \quad \begin{pmatrix} C_{11}(V_3, V_4) & C_{12}(V_3, V_4) \\ C_{21}(V_3, V_4) & C_{22}(V_3, V_4) \end{pmatrix} \frac{d}{dt} \begin{pmatrix} V_3 \\ V_4 \end{pmatrix} = \begin{pmatrix} q_1(V_3, V_4, V_1, V_2) \\ q_2(V_3, V_4, V_1, V_2) \end{pmatrix}$$

where

$$C_{11}(V_3, V_4) = 2 \cdot cgd_P + 2 \cdot cbd_P(V_3, 5.0) + cgd_N + cbd_N(V_3, V_4)$$

$$C_{12}(V_3, V_4) = -cbd_N(V_3, V_4)$$

$$C_{21}(V_3, V_4) = -cbd_N(V_3, V_4)$$

$$(5.2) \quad C_{22}(V_3, V_4) = cgd_N + cbd_N(V_4, 0.0) + cgb_N + cgs_N + cbd_N(V_3, V_4)$$

$$\begin{aligned} q_1(V_3, V_4, V_1, V_2) &= cgd_P \cdot \tfrac{d}{dt}V_1 + (cgd_P + cgd_N) \cdot \tfrac{d}{dt}V_2 - I_N(V_3, V_2, V_4) \\ &\quad - I_P(V_3, V_2, 5.0) - I_P(V_3, V_1, 5.0) \end{aligned}$$

and

$$\begin{aligned} q_2(V_3, V_4, V_1, V_2) &= cgd_N \cdot \tfrac{d}{dt}V_2 + (cgb_N + cgs_N) \cdot \tfrac{d}{dt}V_2 \\ &\quad + I_N(V_3, V_2, V_4) - I_N(V_4, V_1, 0.0). \end{aligned}$$

The nonlinear capacitances are in the form

$$cbd_N(x_1, x_2) = \frac{C_0}{[1 - (x_1 - x_2)]^{\frac{1}{2}}}$$

$$(5.3) \quad cbd_P(x_1, x_2) = \frac{C_0}{[1 - (x_2 - x_1)]^{\frac{1}{2}}}.$$

The remaining capacitors are constant. Functions I_N and I_P represent the drain current for the n-channel and p-channel MOS devices, respectively. The current function I_N depends on the regions of operation of the MOS transistors and it is given by the following expressions

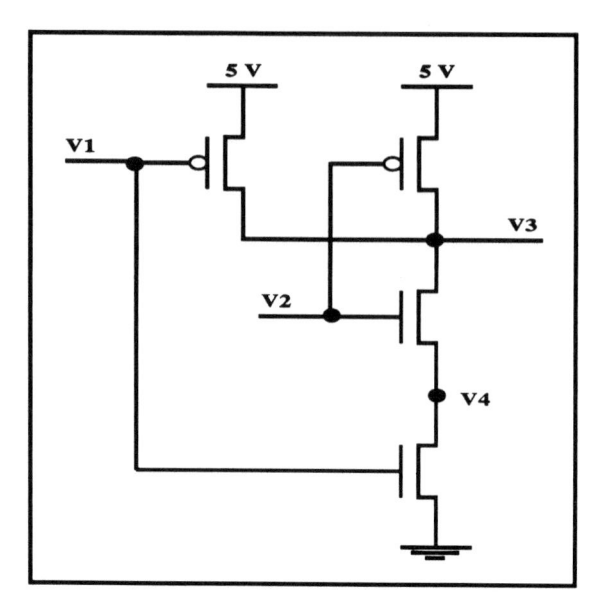

FIG. 1. *Electrical schematics of CMOS NAND gate.*

a) forward region Vds=(Vd-Vs) ≥ 0

$$I_N(Vd, Vg, Vs) =$$

(5.4a) $\begin{cases} 0 & ; \quad Vgs = (Vg - Vs) < Vt = 0.5 \\ \beta Vds(2(Vgs - Vt) - Vds)(1 + \lambda Vds) & ; \quad 0 < Vds < Vgs - Vt \\ \beta(Vgs - Vt)^2(1 + \lambda Vds) & ; \quad Vgs - Vt < Vds \end{cases}$

b) reverse region Vds ≤ 0

$$I_N(Vd, Vg, Vs) =$$

(5.4b) $\begin{cases} 0 & ; \quad Vgd = (Vg - Vd) < Vt \\ \beta Vds(2(Vgs - Vt) + Vds)(1 - \lambda Vds) & ; \quad 0 < -Vds < Vgs - Vt \\ -\beta(Vgs - Vt)^2(1 - \lambda Vds) & ; \quad Vgs - Vt < -Vds \end{cases}$

The p-channel device operates in the same manner as the n-channel device except that all voltages and currents are reversed.

5.2. Results of simulation. The CMOS NAND gate described above was simulated by using the prototype software. The simulation was performed in the time interval $[0.0, 2.3]\mu s$ which was divided into five subintervals (windows) as shown in Fig. 2a and Fig. 3a. The Chebyshev expansion degree was set to 32 for all windows. The results of simulation, voltage V_3, are shown in Fig. 2a together with the driving signals V_1 and V_2. The output shows a logical representation of the NAND function, V_3 is low when V_1 and V_2 are in high level, and V_3 is in high level otherwise. The iteration process is illustrated in Fig. 3b, where the results of some initial iteration steps are shown. The accuracy of the solution was set to $1.0\ mV$ in each window.

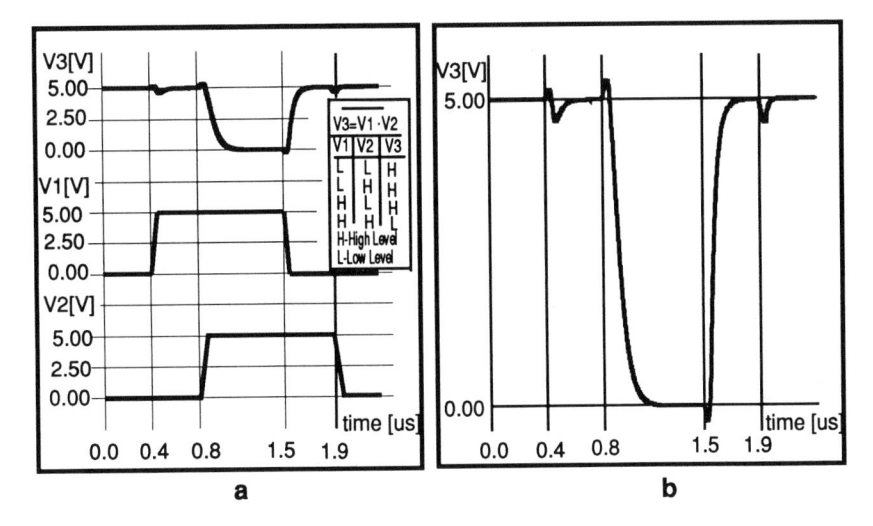

FIG. 2. *Simulation of a NAND gate obtained using SPEC simulator;* **a)** *driving signals, V_1, V_2, and output, V_3;* **b)** *the details of the output voltage V_3.*

6. Conclusions. A set of nonlinear differential equations describing an electronic circuit is linearized with the use of a newly developed linearization scheme. The resulting set of linear differential equations is then transformed to a set of algebraic equations using spectral analysis based on Chebyshev polynomials. Solution of this system is obtained in an efficient way due to the relatively small number of variables and the sparsity of the system. The spectral method has a special importance for bigger circuits where large number of variables is significantly reduced, in comparison with non-spectral methods, reducing a memory requirement and simulation time. The simulations can be successfully performed with the use of smaller computers that don't have large memory. Numerous properties of Chebyshev series [6] bring high potential for further improvements to the simulator.

Acknowledgement. Research described in this paper was supported in part by the NSF grant: MIP-901 7037 and the authors wish to express their gratitude for this support. The idea of writing this paper was conceived during the Summer 1991 Workshop at the Institute of Mathematics and its Applications at the University of Minnesota. O.A. Palusinski extends his thanks to the organizers for the invitation to the workshop and sponsorship. O.A Palusinski was from September 1991 till February 1992 on the sabbatical leave at the University of Karlsruhe supported by a fellowship from the German Science Foundation.

REFERENCES

[1] O.A. Palusinski, F. Szidarovszky, M. Abdennadher, C. Marcjan, K. Reiss, *Accelerated Simulation of Integrated Circuits Using Chebyshev Series* Proceeding of 1992 IEEE-ISCAS.

[2] O.A. Palusinski, M.W. Guarini and S.J. Wright, *Spectral Technique in Electronic Circuit Analysis*, International Journal of Numerical Modeling: Electronic Networks, Devices and Fields, 1, 137-151 (1988).

[3] M.W. Guarini and O.A. Palusinski, *Functional Relaxations and Spectral Techniques in*

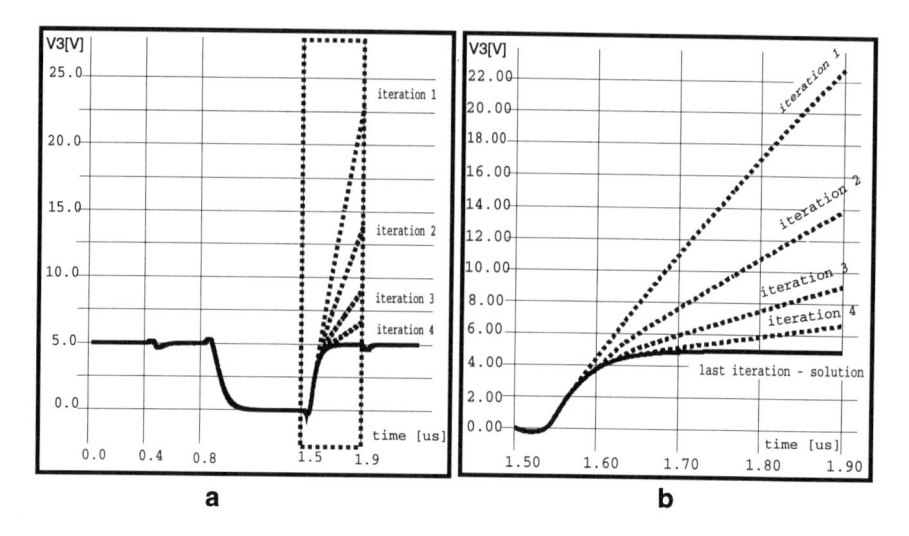

FIG. 3. *Convergence process in computation of transient in the NAND, the example of output variable, V_3;* **a)** *V_3 in the simulation range with marked window for which the iterations were recorded;* **b)** *the details of iteration process in the selected window.*

Computer-Aided Circuit Analysis, International Journal of Numerical Modeling: Electronic Networks, Devices and Fields, **3**, 183-193 (1990).

[4] O.A. Palusinski, M.W. Guarini, *Simulation of MOS Circuits Using Spectral Technique in Relaxation Framework*, COMPEL, the International Journal for Computation and Mathematics in Electrical and Electronic Engineering, vol. 10, No 4, 363-365, Dec, 1991.

[5] L.W. Nagel, *SPICE 2: A Computer Program to Simulate Semiconductor Circuits*, Electronic Research Laboratory Rep. No. ERL-M520, University of California, Berkeley, 1975. Also: SPICE 2 Version 2G6, Department of Electrical Engineering and Computer Science, University of California, Berkeley, ERL Report 1989.

[6] S. Paszkowski, *Numerical Applications of Chebyshev Polynomials and Series*, (in Polish), PWN, Warsaw (1975).

[7] E. Lelarasmee, A.E. Ruehli and A. Sangiovanni-Vincentelli, *The Waveform Relaxation Method for Time Domain Analysis of Large Scale Integrated Circuits*, IEEE Trans. Computer-Aided Des. Integrated Circuits Syst., **CADICS-1** 131-145 (1982).

CONVERGENCE OF WAVEFORM RELAXATION FOR RC CIRCUITS

ALBERT E. RUEHLI[*] & CHARLES A. ZUKOWSKI[†]

Abstract. The waveform relaxation [WR] method of circuit simulation has demonstrated the ability to handle large digital VLSI circuits without sacrificing accuracy. Existing programs use reasonable heuristics for circuit partitioning of present day circuits. As circuit models begin to include more and more parasitic elements, due to shrinking geometries, decreasing signal rise times and increasing operating frequencies, the partitioning will become even more important. This paper addresses the question of how partitioning should be done within complex interconnect models. Specifically, we consider the partitioning of a limiting case RC circuit example, and investigate its convergence properties and optimal time window size.

1. Introduction. Today, the number of parasitic elements required in VLSI circuit models is rapidly increasing due to the decrease in geometries, decrease in signal rise times and the corresponding increase in operating frequencies. Examples include wire resistances and coupling capacitances. WR (waveform relaxation) [1] programs for circuit simulation have demonstrated their ability to handle today's large digital VLSI circuits with accuracy similar to that of SPICE [2] or ASTAP [3]. Little work has been done so far on WR programs which efficiently include models with large numbers of parasitic elements. Hence, handling circuits in a WR program with an increased number of parasitic elements is a problem of high priority.

In this paper we consider the special case of the low pass RC circuit shown in Fig. 1 which is typical of a simplified model for the interconnects in VLSI circuits. We chose this circuit, with no forced drive connection, since it represents the most difficult case for WR if it is partitioned at the resistor. This circuit is a complementary special circuit to a so called "high pass" circuit which has been analyzed before [4]. The circuit shown in Fig. 1 cannot be partitioned with the diagonally dominant Norton algorithm [5] which is used in several WR programs. Waveform Relaxation for RC circuits has been considered two recent papers [6, 7]. Further, relaxation for RC circuits has been investigated earlier in the context of bounding but the properties of the convergence were not analyzed [8].

Here we are specifically interested in convergence behavior for the RC circuit. First, it is evident that the convergence for low frequencies is in question since the capacitors are open circuits at *dc*. Fortunately, we can show that the circuit converges, as expected from the basic WR convergence proof. Further, the rate of convergence is of importance. For efficiency reasons we hope that the relaxation converges in less than five iterations. In fact, we derive a measure of the time window over which rapid convergence can be expected.

2. Analysis of RC circuit. The low pass RC circuit in Fig. 1 is a common model for VLSI interconnects. In a sense it represents a worst case situation since any other circuit element connected from either node to ground will improve local convergence. The WR equations for this circuit when partitioned through the resistor

---

[*] IBM T. J. Watson Research Center, Yorktown Heights, NY 10598
[†] Columbia University, New York, NY 10027-6699

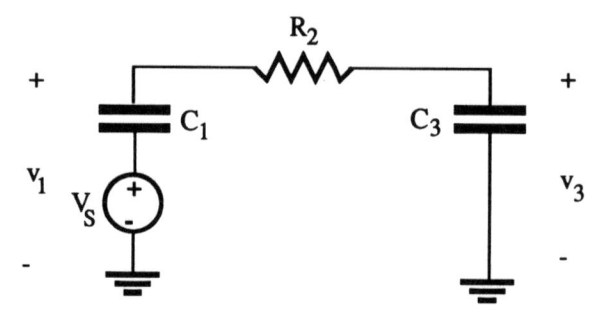

FIG. 1. Low pass circuit

are

(1) $\quad \dot{v}_1^{(k)}(t) + \alpha v_1^{(k)}(t) = \alpha v_3^{(k-1)}(t) + \dot{V}_s(t)$

(2) $\quad \dot{v}_3^{(k)}(t) + \beta v_3^{(k)}(t) = \beta v_1^{(k-1)}(t)$

where $\alpha := 1/R_2C_1$ and $\beta := 1/R_2C_3$.

We can analyze the circuit in the frequency domain by using the Laplace transform for the homogeneous equation or

(3) $$\begin{bmatrix} (s+\alpha) & 0 \\ 0 & (s+\beta) \end{bmatrix} \begin{bmatrix} \tilde{v}_1^{(k)} \\ \tilde{v}_3^{(k)} \end{bmatrix} = \begin{bmatrix} 0 & \alpha \\ \beta & 0 \end{bmatrix} \begin{bmatrix} \tilde{v}_1^{(k-1)} \\ \tilde{v}_3^{(k-1)} \end{bmatrix}$$

where $\tilde{v}(s)$ is the Laplace transform of $v(t)$. The above equation is, in matrix notation,

(4) $$(s\mathbf{I} + \mathbf{M})\tilde{\mathbf{v}}^{(k)} = \mathbf{N}\tilde{\mathbf{v}}^{(k-1)},$$

where the definitions of $\mathbf{M}$ and $\mathbf{N}$ are evident from comparisons of the last two equations. We can rewrite Eq. 4 as

(5) $$\tilde{\mathbf{v}}^{(k)}(s) = K(s)\tilde{\mathbf{v}}^{(k-1)}(s).$$

We utilize the following theorem [9] to show that the WR iteration has a potential problem.

THEOREM 2.1. *Let $\chi = L^p(\mathbf{R}_+, C^n)$ with $1 \leq p \leq \infty$ and assume that the eigenvalues of $\mathbf{M}$ have positive real parts. Then the spectral radius of the symbol $K(s)$ is*

(6) $$\rho(K) = max_{\omega \in \mathbf{R}} \rho((i\omega \mathbf{I} + \mathbf{M})^{-1}\mathbf{N}) \quad \square$$

Evaluating $\rho(K)$ we find

$$K(s) = \begin{bmatrix} 0 & \frac{\alpha}{s+\alpha} \\ \frac{\beta}{s+\beta} & 0 \end{bmatrix} \tag{7}$$

and it is clear that the maximum occurs for $\omega = 0$ where $\rho(K(0)) = 1$, which indicates that convergence problems occur at $\omega \to 0$. Fortunately we show that this does not imply that there is a problem over any finite time interval, but only indicates that convergence is non-uniform and degrades over the infinite time interval. Also, this difficulty does not exist if other parasitic elements are connected at either node.

We simplify the problem by choosing $\alpha = \beta = 1$ without any loss of key insight. By executing the WR interation in the s-domain, we arrive at the solution for iteration (k):

$$\tilde{v}_1^{(k)}(s) = \sum_{m=1}^{(k)} \frac{1}{(s+1)^{2m-1}}. \tag{8}$$

The solution in the time domain corresponding to Eq. 8 is found from the inverse Laplace transform as

$$v_1^{(k)}(t) = e^{-t} \sum_{m=0}^{(k-1)} \frac{t^{2m}}{(2m)!}. \tag{9}$$

Taking the limit as the iteration index $k \to \infty$ we find that the limit is

$$v_1(t) = e^{-t}\cosh(t). \tag{10}$$

To verify that the limit is indeed the exact solution, we start by finding the s-domain response of the circuit in Fig. 1,

$$\tilde{v}_1 = \frac{(s+\beta)}{s(s+(\alpha+\beta))}, \tag{11}$$

for a unit initial voltage. Converting this into its corresponding time solution, we find

$$v_1(t) = \frac{1}{2}(1 + e^{-2t})u(t), \tag{12}$$

which matches the limit waveform.

CRC CIRCUIT

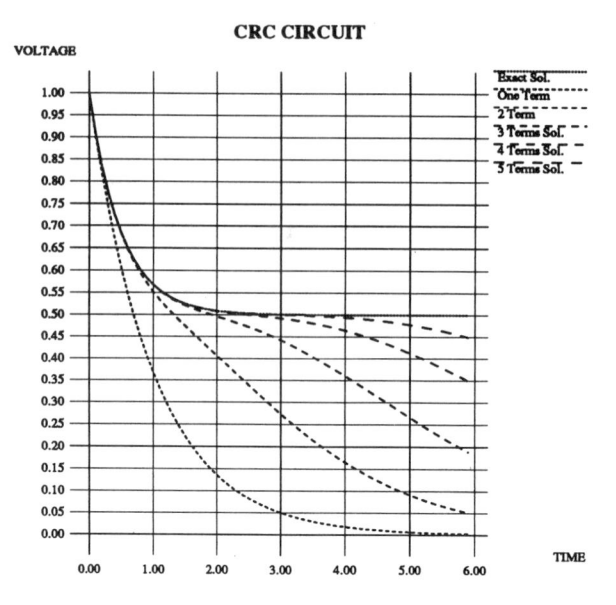

VOLTAGE

```
                                                    Exact Sol. ........
1.00                                                One Term  ........
0.95                                                2 Term    - - - -
0.90                                                3 Terms Sol. -- -
0.85                                                4 Terms Sol. —  —
0.80                                                5 Terms Sol. ——
0.75
0.70
0.65
0.60
0.55
0.50
0.45
0.40
0.35
0.30
0.25
0.20
0.15
0.10
0.05
0.00
    0.00   1.00   2.00   3.00   4.00   5.00   6.00   TIME
```

FIG. 2. *Convergence of initial iterations*

3. Convergence results. The above analysis indicates that a reasonable WR sequence exists even for this *worst case* RC circuit. The questions which are addressed in this section are the convergence rate and the choice of time window size T.

A sufficient condition for the convergence of WR sequences for conventional circuits is that a capacitor is connected to ground from each node [1]. Furthermore, RC circuits that are driven generally exhibit uniform convergence over the infinite time interval. For this *worst case* circuit, we can show that the convergence does not exhibit such uniformity because the sequence does not converge at $t = \infty$. We start by the applying the final value theorem $\lim_{\sigma \to 0} \sigma \hat{f}(\sigma) = f(\infty)$ to Eq. 11 to find that the response of the RC circuit at $t \to \infty$ is given by

$$(13) \qquad v_1(\infty) = \frac{\beta}{\alpha + \beta}.$$

which evaluates to $v_1(\infty) = \frac{1}{2}$ for $\alpha = \beta = 1$. But for $t \to \infty$ where $\dot{V}_s(t) = 0$ the WR sequence degenerates to $v_1^{(k)} = v_1^{(k-1)}$. If the initial waveform approaches anything other than $\frac{1}{2}$ for $t \to \infty$, the sequence will never get any closer at this limit. This non-uniformity of convergence implies that accuracy is only reached gradually for larger and larger times, and implies that time windowing is appropriate.

Next, we investigate the rate of convergence for the model circuit. One of the issues of interest is the appropriate selection for time window size for this class of circuits. Most WR programs allow a time window size T that is less than the entire analysis time for increased efficiency. As is typical, a tradeoff exists in our example between the number of iterations and the time window, and this relationship is quantified in the following lemma.

TABLE 1
Window times

Iteration	Time Windows
1	0.74
2	1.47
3	2.21
4	2.94
5	3.68

LEMMA 3.1. *Let $T \in \mathbf{R}_+$, then the WR sequence converges rapidly after the k-th iteration for $k \geq \frac{eT}{2}$.* □

The proof is fairly straight forward using the approximate identity $(2m)! \cong \sqrt{2\pi}(2m)^{2m+\frac{1}{2}}e^{-2m}$. After some algebraic manipulations this leads to

$$\text{Error}(v_1^{(k)}(t)) = \frac{e^{-t}}{2\sqrt{\pi}} \sum_{m=k}^{\infty} \left(\frac{et}{2m}\right)^{2m} \frac{1}{\sqrt{m}} \tag{14}$$

In fact, we can see easily from Eq. 14 that the coefficients decrease faster than $O(\frac{1}{k^{2k}})$ for the conditions on the time T and the iteration index k given in the lemma. Fig. 2 shows how the first five iterations converge. In fact, the time window T of rapid convergence is clearly visible. We can conclude from the above condition that a reasonable choice for a time window as a function of iteration count is $k > \frac{eT}{2}$, as pictured. For a comparison, we give the values for the equality in Table 1.

We do get an indication from Fig. 2 and Table 1 how the useful time window grows with the number of iterations. Of course, with our choices of $\alpha = \beta = 1$, this is normalized to unit time constants. This implies, in general, that time constants associated with this subcircuit should be sufficiently large such that the time window is large enough that it does not constrain the number of time steps in a window too much. "Large enough" may mean that the numerical integration needs at least 10 time points in a particular window. This condition may be guaranteed for a transition or spike in the waveform. However, a problem with window size may exist if the WR code has the same global time windows for the entire circuit.

4. Conclusions. The work presented in this paper increases understanding of WR partitioning for an extended class of circuits. A simple example that captures some of the properties of large interconnect subcircuits was analyzed in detail, and the rate of convergence was related to time window size. We conclude that large interconnect subcircuits can be partitioned efficiently if appropriate care is taken in computing local feedback and choosing time window sizes.

REFERENCES

[1] E. Lelarasmee, A. E. Ruehli, and A. L. Sangiovanni-Vincentelli. The waveform relaxation method for the time-domain analysis of large scale integrated circuits. *IEEE Trans. on CAD of ICs and Systems*, CAD-1(3):131-145, July 1982.

[2] L. W. Nagel. SPICE2, a computer program to simulate semiconductor circuits. Memo UCB/ERL M520, University of California, Berkeley, May 1975.

[3] W. T. Weeks, A. J. Jimenez, G. W. Mahoney, D. Mehta, H. Quassemzadeh, and T. R. Scott. Algorithms for ASTAP - a network analysis program. *IEEE Trans. on Circuit Theory*, CT(20):628-634, November 1973.

[4] U. Miekkala, O. Nevanlinna, and A. E. Ruehli. Convergence and circuit partitioning aspects for waveform relaxation. *Proc. of Fifth Distrib. Memory Computing Conf.*, D. W. Walker and Q.F. Strout, Eds., IEEE Comp. Society Press, pages 605-611, 1990.

[5] J. White and A. L. Sangiovanni-Vincentelli. Partitioning algorithms and parallel implementations of waveform relaxation algorithms for circuit simulation. *IEEE Proc. Int. Symp. Circuits and Systems, (ISCAS-85)*, pages 1069-1072, 1985.

[6] A. E. Ruehli, G. Gristede, and C. Zukowski. On partitioning and windowing for waveform relaxation. In *Proc. Seventh Int. Conf. Numerical Analysis of Semiconductor Devices and Circuits*, pages 69-72, Boulder, Colorado, April 1991. Front Range Press.

[7] B. Leimkuhler, U. Miekkala, and O. Nevanlinna. Waveform relaxation for linear RC circuits. *Impact of Comp. in Science and Engineering*, pages 123-145, 1991.

[8] C. Zukowski. Relaxing bounds for linear RC mesh circuits. *IEEE Trans. on CAD of ICs and Systems*, pages 305-312, April 1986.

[9] U. Miekkala and O. Nevanlinna. Convergence of waveform relaxation method. *IEEE Int. Conf. Circuits and Systems (ISCAS-88)*, pages 1643-1646, 1988.

SWITCHED NETWORKS

J. VLACH[*] AND D. BEDROSIAN[**]

Abstract. Inconsistent initial conditions, which can exist in switched networks, cannot be handled by the usual integration routines. A method based on numerical inversion of the Laplace transform was developed. It is equivalent to a high-order integration and can handle inconsistent initial conditions, discontinuous functions and Dirac impulses. The method was used to write programs for analysis of switched networks.

1. Introduction. The use of switched networks became possible with the development of semiconductor switches which can operate at high speed and handle large voltages and currents. Analysis of networks with such switches is possible by using complex semiconductor models and classical simulators, but computer time may become excessive while the exact switching responses are rarely of interest. Making the switches ideal, by assuming their state to be either a true open or short circuit, reduces considerably computing times, but new problems are created. The most difficult one is the possibility of inconsistent initial conditions.

The easiest way to visualize inconsistent initial conditions is to consider two capacitors, one of them discharged and one of them charged to some voltage. If the two capacitors are suddenly connected by an ideal switch, the charged capacitor immediately transfers part of its charge to the other capacitor and their common voltage at the instant of switching becomes equal. The charge transfer is achieved by an infinitely short impulse of current, the so called Dirac impulse. Strictly speaking, it is not a function, but is is used widely in electrical engineering. It is denoted by $\delta(t)$ and can be defined as a rectangular impulse having duration T and height $1/T$, with $\lim T \to 0$.

The simple problem described above is easy to resolve, but in more complicated cases it is very difficult to find the initial conditions after switching and very often the possibility of impulsive currents is simply disregarded. This can have detrimental influence on the results.

This paper describes development of an integration method which automatically takes care of inconsistent initial conditions, discontinuous functions or Dirac impulses at the instant of switching. Only small examples and brief description are given here. For details, the reader is referred to the references.

2. Development of the method. It is known that the Laplace transform correctly handles Dirac impulses and inconsistent initial conditions, but its use in linear networks requires the knowledge of the eigenvalues, a step which is impractical for larger systems. We avoid their evaluation by developing a numerical method of inversion. It is applied like the commonly known integration methods, but it does

---

[*]Department of Electrical and Computer Engineering, University of Waterloo, Waterloo, Ontario, Canada.
[**]Analogy, Beaverton, Oregon 97075 USA.

retain the ability of the Laplace transform to correctly handle inconsistent initial conditions. Consider the Laplace transform equation

$$\nu(t) = \frac{1}{2\pi j} \int_{c-j\infty}^{c+j\infty} V(s)e^{st}ds.$$

and use the substitution $s = \frac{z}{t}$ to obtain

$$\nu(t) = \frac{1}{2\pi j t} \int_{\hat{c}-j\infty}^{\hat{c}+j\infty} V\left(\frac{z}{t}\right) e^z dz.$$

Approximate e^z with the Pade rational function:

$$e^z \approx R_{N,M}(z) = \frac{\sum_{i=0}^{N}(M+N-i)!\binom{N}{i}z^i}{\sum_{i=0}^{M}(-1)^i(M+N-i)!\binom{M}{i}z^i}.$$

For $N < M$ it is equivalent to

$$R_{N,M}(z) = \sum_{i=1}^{M} \frac{K_i}{z - z_i},$$

where the numerical evaluation of the roots, z_i, and residues, K_i, is done only once. The inversion formula becomes

$$\nu(t) \approx \hat{\nu}(t) = \frac{1}{2\pi j t} \sum_{i=1}^{M} \int_{\hat{c}-j\infty}^{\hat{c}+j\infty} \frac{K_i}{z - z_i} V\left(\frac{z}{t}\right) dz.$$

Applying the residue calculus for the integral we arrive at

$$\hat{\nu}(t) = -\frac{1}{t} \sum_{i=1}^{M} K_i V\left(\frac{z_i}{t}\right)$$

It is proved in [1] (Appendix C) that the integration by residues is possible provided the overall integrated function $V\left(\frac{z}{t}\right) R_{N,M}^{(z)}$ has two more poles than zeros. Since realizable functions have at most one more zero than pole, the approximation $M = 4$ and $N = 0$ will suffice, but other choices are valid as well. If M is an even integer, we can simplify the formula to

(1) $$\hat{\nu}(t) = -\frac{2}{t} \sum_{i=1}^{M/2} \text{Re}\left[K_i V\left(\frac{z_i}{t}\right)\right]$$

and consider only upper half plane poles and their residues. It is shown in [1] that formula (1) approximates the first $M+N+1$ terms of the Taylor expansion of $\nu(t)$ for any $t > 0$.

In the case of networks we can use the system equations

(2) $$(\mathbf{G} + s\mathbf{C})\mathbf{V} = \mathbf{W}(s) + \mathbf{I}(0^-)$$

Here $\mathbf{G}$ and $\mathbf{C}$ are real matrices, $\mathbf{W}$ is a vector of external sources, $\mathbf{V}$ is the solution vector in the Laplace domain and $\mathbf{I}$ is a vector of initial capacitor voltages and initial inductor currents. In this equation, every s is substituted by z_i/t, the evaluation is repeated $M/2$ times, and the results are used in (1) to obtain the time domain vector, $\mathbf{v}(t)$. The initial conditions vector, $\mathbf{I}$, is obtained as the product

$$\mathbf{I}(0^-) = \mathbf{C}\mathbf{v}(t)$$

where $\mathbf{v}(t)$ is the result from the previous step. The possibility of resetting the initial conditions by means of initial voltages and currents makes the procedure equivalent to a numerical integration formula, somewhat resembling the Runge-Kutta methods, because previous results are not needed. However, evaluations are done in complex. In the following we assume the selection $M = 4$, $N = 0$.

3. Initial conditions. When dealing with inconsistent initial conditions, we may need the answer to one of two problems:

1. Find the initial condition at $t = 0^+$.
2. Establish that the Dirac impulse exists at $t = 0$ and find its area.

Equation (1) cannot evaluate the response at $t = 0$ due to the division by t, but a choice of a small t could approximate the initial conditions at $t = 0^+$. Such approximation turns out to be acceptable in some situations and unacceptable in others. To resolve the problem consider the network in Figure 1(a). The input is a Dirac impulse, the output in the Laplace domain is

$$V = \frac{1}{s+1}$$

and in the time domain

$$\nu(t) = e^{-t}.$$

Application of formula (1) gives a value which is correct to about 15 decimal digits (on a 16 decimal digit machine) for every t in the interval from 10^{-12} to 10^{-3}, see Figure 2(a). In each case the error corresponds to a *single* step. The error starts growing for larger t, but that is to be expected, since the formula is an approximation. In this case the Dirac impulse *does not* appear at the output.

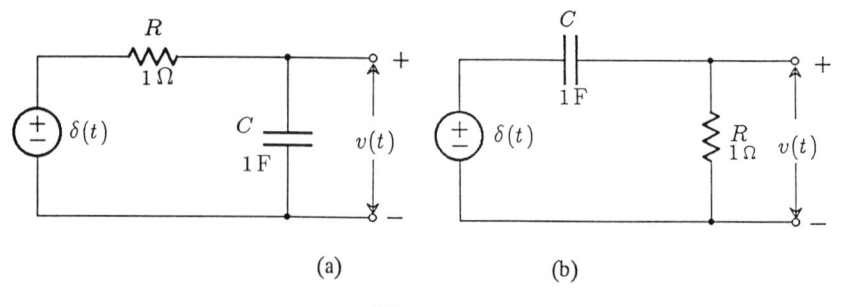

[Figure 1 showing two circuits labeled (a) and (b)]

(a)   (b)

Figure 1.

The situation changes drastically for Figure 1(b). It is the same network but the output is taken at a different node. In the Laplace domain the output is

(3) $$V(s) = \frac{s}{s+1} = 1 - \frac{1}{s+1}$$

and in time domain

$$\nu(t) = \delta(t) - e^{-t}$$

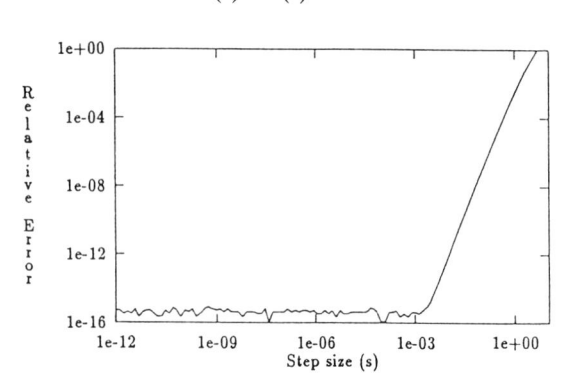

[Figure 2(a): plot of Relative Error vs Step size (s)]

Figure 2(a).

A Dirac impulse appears at the output. In this case formula (1) gives a large error for a small step, $\varepsilon = 10^{-4}$ for $t = 10^{-12}$, see Figure 2(b), again for a single step. However, the error decreases almost linearly to $\varepsilon \approx 10^{-13}$ for $t \approx 10^{-3}$. Here the integration error is acceptable, but the solution is a poor approximation of the initial condition at $t = 0^+$.

In order to get correct initial conditions at $t = 0^+$, even in the presence of the Dirac impulse at $t = 0$, we propose to first make one large step forward, to get to the minimum of the integration error. Afterwards, starting from this new point, we make an exactly equal step backward. This backward step is essentially error-free, since there is no impulse at the new starting point. The error in this step is the same as in Figure 2(a). As a result of this two-step procedure we get correct initial condition at $t = 0^+$.

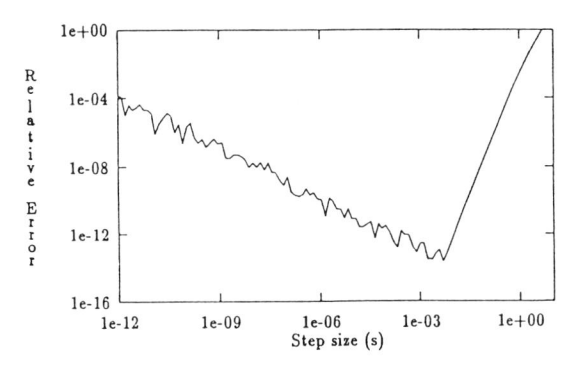

[Figure: Relative Error vs Step size (s), log-log plot from 1e-12 to 1e+00]

Figure 2(b).

An explanation of why there is such a difference in the results is also available. Consider the network in Figure 1(b) for which the output is

$$V(s) = \frac{s}{s+1} = 1 - \frac{1}{s+1}.$$

At $t = 0^+$ the correct solution is $\nu(0^+) = -1$. If this function is inserted into formula (1), we obtain

$$\hat{\nu}(t) = -\frac{2}{t} \operatorname{Re}\left[K_1\left(1 - \frac{t}{z_1+t}\right) + K_2\left(1 - \frac{t}{z_2+t}\right)\right].$$

For very small t and finite arithmetic precision the fractions will be dominated by the units. The expression effectively reduces to

$$\hat{\nu}(t) = -\frac{2}{t} \operatorname{Re}[K_1 + K_2] = 0$$

due to the fact that the real parts of the residues are equal in magnitude but have opposite signs. This error is eliminated by the two-step method.

4. Representation of $\delta(t)$. If the network contains switches whose state depends on the solution variables of the problem, it may be important to discover whether a Dirac impulse does or does not appear at the instant of switching. For the purpose of explanation consider the network in Figure 3 where the transistor Q_1 is switched externally by a square wave. Both the transistor and the diode are modeled as ideal switches. When the external square wave causes the transistor to act as a short circuit, the voltage at the upper end of the inductor is negative and the diode does not conduct. The current flowing through the inductor builds around it a magnetic field. When the transistor switch is suddenly opened by the external square wave, the flow of current is interrupted and, due to the magnetic field, a positive Dirac impulse will appear at the upper end of L. This impulse closes the diode and the current through the inductor can continue to flow into the right part of the network.

The sequence of these events must be discovered by the analysis method in order to correctly handle the switching. Thus we must discover whether there is or is not an impulse at the instant of switching. The Dirac impulse cannot be represented in the computer, because it has zero duration and is infinitely large. However, we can represent the solution at the instant of switching by two components,

(4) $$\nu(0) = \nu(0^+) + \nu_\delta \delta(t).$$

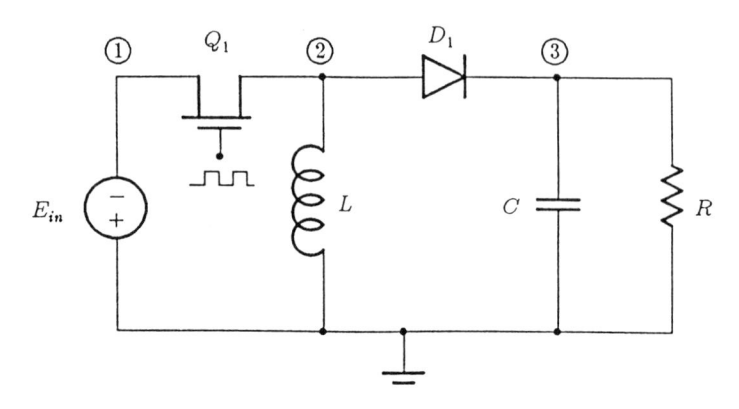

Figure 3. Switched network.

The first one is the true initial condition after switching, obtained by the two-step method explained above. The term ν_δ is a multiplicative factor corresponding to the area of the Dirac impulse. This term is always finite, can be stored in the memory of the computer, and has zero value when no impulse has occurred at the instant of switching. The problem is to find this coefficient.

Consider the situation in which we have reached the instant $t = 0^-$, just before switching. Using the two step method we can obtain the term $\nu(0^+)$ but we still do not know whether the impulse has occurred at $t = 0$. To discover its existence we can calculate the area between $t = 0^-$ and $t = 0^+$ by the same two-step method. Since in the Laplace domain the integration is expressed by division by s, we evaluate

$$\int_{0^-}^{t} \nu(\tau)d\tau = -\frac{2}{t}\sum_{i=1}^{2} \text{Re}\left[K_i\left(\frac{t}{z_i}\right)V\left(\frac{z_i}{t}\right)\right]$$

and do the same for the step back. The difference of the two areas is the area of the Dirac impulse. Note that this integration is done almost for free; it represents only four additional multiplications for our selection of $M = 4$ and $N = 0$, and for both steps forward and back.

It is interesting to see the accuracy of this method. For the function

$$V(s) = -5 + \frac{3}{s+2}$$

whose time domain response is

$$\nu(t) = -5\delta(t) + 3e^{-2t},$$

the forward step with 5 ms length provides

$$\int_{0^-}^{0.005} \nu(t)dt = -4.98850758$$

with relative error 2.5×10^{-13}. The backward step provides

$$\int_{0.005}^{0^+} \nu(t)dt = -0.01492525$$

with relative error 8.4×10^{-11}. The sum of the two integrals is -5.00000, with relative error 3.7×10^{-15}. We thus have an accurate method to find out whether a Dirac impulse did or did not appear at the instant of switching.

5. **Application.** The method was used to write programs for analysis of switched networks driven by external clocks or operated by internally controlled switches. If only external periodic clocks are present, frequency domain analysis is possible [2,3]. If the network has some internally controlled switches, then only time domain analysis can be used [4,5]. However, if such a network is periodic, then one more application is possible, an accelerated method for finding the steady state [6].

Steady state can always be reached by integrating the network equations over sufficiently long time, until the transients die out. This is usually very expensive and the uncertainty always remains whether the steady state has actually been reached with sufficient accuracy. An accelerated method was developed, based on the idea that in steady state the initial conditions at the beginning of the period must be equal to the final conditions at the end of the same period. An error function can be defined

(5) $$E[\mathbf{v}(0^-)] = \mathbf{v}(T^-) - \mathbf{v}(0^-)$$

and an iterative method used to reduce the error to zero. A suitable method is the Newton-Raphson iteration, based on the equations

$$J^{(k)}[\mathbf{v}(0^-)]\Delta\mathbf{v}^{(k)} = -E[\mathbf{v}(0^-)]$$
$$\mathbf{v}^{(k+1)} = \mathbf{v}^{(k)} + \Delta\mathbf{v}^{(k)}$$

where the Jacobian matrix is

$$J^{(k)}[\mathbf{v}(0^-)] = \frac{\partial E[\mathbf{v}(0^-)]}{\partial \mathbf{v}(0^-)} = \frac{\partial \mathbf{v}(T^-)}{\partial \mathbf{v}(0^-)} - I$$

The integration method explained above is used to overcome problems with switching and possible inconsistent initial conditions. It is also used in the evaluation of the Jacobian. Mathematical details are given in [6,7] where the efficiency of the method is demonstrated on several practical examples.

REFERENCES

[1] J. VLACH AND K. SINGHAL, *Computer Methods for Circuit Analysis and Design*, Van Nostrand Reinhold, New York, 1983.

[2] A. OPAL AND J. VLACH, *Consistent initial conditions of linear switched networks*, IEEE Transactions on Circuits and Systems, CAS-37 (3), March, 1990, pp. 364–372.

[3] A. OPAL AND J. VLACH, *Analysis and sensitivity of periodically switched linear networks*, IEEE Transactions on Circuits and Systems, CAS-36 (4), April, 1989, pp. 522–532.

[4] D. BEDROSIAN AND J. VLACH, *Time-Domain Analysis of Networks with Internally Controlled Switches*, Vol. 39 (3), March, 1992, pp. 199–212.

[5] D. BEDROSIAN AND J. VLACH, *Analysis of Switched Networks*, International Journal of Circuit Theory and Applications, Vol. 20 (3), May-June, 1992, pp. 309–325.

[6] D. BEDROSIAN AND J. VLACH, *Accelerated Steady-State Method for Networks with Internally Controlled Switches*, IEEE Transactions on Circuits and Systems I: Fundamental Theory and Applications, July 1992, Volume 39, Number 7, pp. 520–530.

[7] D. BEDROSIAN AND J. VLACH, *An Accelerated Steady-State Method for Networks with Internally Controlled Switches*, IEEE International Conf. on Computer-Aided Design, Santa Clara, California, November 11-14, 1991, pp. 24–27.